T0074871

Performance of DFIG and PMSG Wind Turbines

Due to environmental pollution and climate change, the use of renewable energy sources as an alternative means of power generation is on the rise globally. This is because of their clean nature, which makes them eco-friendly with little or no pollution compared to traditional fossil fuel power-generated power plants.

Among the various renewable energy sources, wind energy is one of the most widely employed, due to its promising technology. Wind turbine technologies could be classified into two groups as follows: Fixed Speed Wind Turbines (FSWTs) and Variable Speed Wind Turbines (VSWTs). There have been tremendous improvement in wind turbine technology over the years, from FSWTs to VSWTs, as a result of fast innovations and advanced developments in power electronics. Thus, VSWTs have better wind energy capture and conversion efficiencies, less acoustic noise and mechanical stress, and better power quality in power grids without support from external reactive power compensators due to the stochastic nature of wind energy.

The two most widely employed VSWTs in wind farm development are the Doubly Fed Induction Generator (DFIG) and the Permanent Magnet Synchronous Generator (PMSG) wind turbines. In order to solve transient stability intricacies during power grid faults, this book proposes different control strategies for the DFIG and PMSG wind turbines.

Performance of DFIG and PMSG Wind Turbines

Kenneth E. Okedu

CRC Press
Taylor & Francis Group
Boca Raton London New York

CRC Press is an imprint of the
Taylor & Francis Group, an **informa** business

First edition published 2023
by CRC Press
6000 Broken Sound Parkway NW, Suite 300, Boca Raton, FL 33487-2742

and by CRC Press
4 Park Square, Milton Park, Abingdon, Oxon, OX14 4RN

CRC Press is an imprint of Taylor & Francis Group, LLC

ISBN: 978-1-032-39507-4 (hbk)
ISBN: 978-1-032-39690-3 (pbk)
ISBN: 978-1-003-35091-0 (ebk)

DOI: 10.1201/9781003350910

Typeset in Times
by SPi Technologies India Pvt Ltd (Straive)

Contents

Preface

With the recent proliferation and penetration of wind farms into existing power grids, it is paramount to conduct numerous studies to counter grid disturbances based on operational grid codes. Electric power from wind energy could be extracted by employing the promising technologies of the Doubly Fed Induction Generator (DFIG) and Permanent Magnet Synchronous Generators (PMSG) wind turbines. In order to solve transient stability intricacies posed by the stochastic nature of wind energy during grid faults, this book would propose different control strategies for DFIG and PMSG wind turbines. The control strategies would be based on Fault Current Limiters (FCLs). The Series Dynamic Braking Resistor (SDBR) would be the first FCL to be investigated in this book. The best location to place the SDBR on the machine side and grid side converters of both DFIG and PMSG wind turbines would be investigated, considering different switching strategies. Efforts would also be made to determine the suitable sizing of the SDBR. The performance of the SDBR would be investigated at various network strengths in weak and strong grids for both wind turbines. Both severe three-line-to-ground faults and asymmetrical faults of line-to-line, double-line-to-ground and one-line-to-ground would be used to test the robustness and rigidity of the controllers of the wind generators.

In weak grids, the challenges of network stability are a result of wind energy penetration. The Voltage Source Inverter (VSI) based on Pulse Width Modulation (PWM) is employed widely in interfacing sources with regard to renewable energy and the grid. The utilization of these inverters would cause stability issues in the power grid. The studies carried out in the literature show that VSI control could affect the stability of power grids. In addition, the stability of grid-connected VSI can be affected by a weak grid. A grid that has a low Short Circuit Ratio (SCR) is said to be weak. In other words, a grid is said to be weak if it has an impedance that is high and low inertia constant. In this book, the performance of both wind turbines would be tested in weak, normal and strong grids, in addition to the SDBR implementation on both wind turbines. The recently stipulated grid codes require that wind generators re-initiate normal power production after grid voltage sag. This book will also present a comparative performance of two commonly employed variable speed wind turbines in today's electricity market: the DFIG and the PMSG wind turbines. The evaluation of both wind turbines was done for weak, normal and strong grids, considering the same machine ratings of the wind turbines. Because of the critical situations of the wind turbines during faulty conditions in the weak grids, an analysis was done considering the use of effective SDBR for both wind turbines. The grid voltage variable was employed as the signal for switching the SDBR in both wind turbines during transient state. Also, an overvoltage protection system was considered for both wind turbines using the DC chopper in the DC-link excitation circuitry of both wind turbines. Furthermore, a combination of the SDBR and DC chopper was employed in both wind turbines at weak grid condition in order to improve the performance of the variable speed wind turbines.

The performance of other FCLs would be investigated in the DFIG and PMSG wind turbines in this book. The FCLs are the Bridge FCL and the Capacitive FCL. Controlling variable speed wind turbines during transient state is challenging. The use of variable speed wind turbines based on PMSG is on the rise due to some of the features of the wind turbine. According to the grid code requirements, grid-connected wind turbine systems should achieve active power control and provide Low Voltage Ride Through (LVRT) capability. Thus, the primary target of the wind turbine control system is to keep the turbine connected to the grid during grid disturbances or failures. In this book, the SDBR and the Bridge Fault Current Limiter (BFCL) were used to improve the LVRT of PMSG wind turbines. The topology of the PMSG grid side voltage source converter, with the SDBR and BFCL, was modeled during steady and transient states. The performance of both schemes on the PMSG was analyzed and compared during a severe balanced fault scenario. In addition, a scenario with any of the schemes was also considered. For fair comparison, the PMSG wind turbine was operating at its rated speed during the low voltage and the same conditions of operation were used for all the considered scenarios.

This book proposes a supercapacitor strategy for improving the capability of grid-connected Doubly Fed Induction Generator (DFIG) wind turbines during fault scenarios. Supercapacitors are one of the important components in sustainable energy systems that are commonly used to store energy. In DFIGs, the super-capacitor is used to compensate for voltage dips and damping oscillations. In this book, a new topology of supercapacitor system was investigated in a DFIG wind turbine during transient state. The model system employed was a DFIG connected to the earlier wind turbine technology of a fixed speed squirrel cage induction generator. Efforts were made to determine the effective parameters and switching strategies of the supercapacitor by considering different scenarios, in order to improve the transient state of the wind generator. The results obtained under severe grid fault were compared considering the different parameters of the resistance, inductance and capacitance of the supercapacitor. The DC-link voltage and grid voltage switching strategies of the supercapacitor were investigated. Furthermore, the results of the proposed DFIG supercapacitor were compared with the traditional parallel capacitor scheme for DFIG system. For fair comparison between the DFIG supercapacitor and parallel capacitor-based solution, the capacitance value considered was the same to buffer the transient energy.

This book also investigates the transient performance of the two DFIG and the PMSG wind turbines. The machine parameters of both wind turbines were varied considering different scenarios, while keeping other parameters constant, in order to study the behavior of the wind generators. The study was carried out using the same operating conditions of rated wind speed, based on the wind turbine characteristics of both wind turbine technologies. The wind turbines were subjected to a severe three-phase-to-ground bolted fault, in order to test the robustness of their controllers during grid fault conditions. Efforts were made to carry out an extensive comparative study to investigate the machine parameters that have more influence on the stability of the different wind turbines considered in this study. Effective machine parameter selection could help solve Fault Ride Through (FRT) problems in a cost-effective way for both VSWTs, without considering external circuitry and changing of the original architecture of the wind turbines.

The system performance carried out in this book was evaluated using Power System Computer Design and Electromagnetic Transient Including DC (PSCAD/EMTDC) platform. The same conditions of operation were used in investigating the various scenarios considered in this study, for effective comparison.

Kenneth E. Okedu,
Visiting Professor, Department of Electrical and
Communication Engineering,
National University of Science and Technology,
Muscat, Sultanate of Oman.
Adjunct Professor, Department of Electrical
and Electronic Engineering,
Nisantasi University,
Istanbul, Turkey.

Acknowledgments

The author of the book would like to thank God Almighty, his wife (Imonina Blessing Oked-Kenneth), daughter (Imonisa Ambless Okedu-Kenneth), parents (Sir and Mrs. Simon Williams Okedu) and mentors (Prof. Junji Tamura and Prof. S. M. Muyeen) for their support, love and care.

Author

Kenneth E. Okedu was a research fellow in the Department of Electrical and Computer Engineering, Massachusetts Institute of Technology (MIT), Boston, USA, in 2013. He obtained his PhD from the Department of Electrical and Electronic Engineering, Kitami Institute of Technology, Japan, in 2012. He received his BSc and MEng in Electrical and Electronic Engineering from the University of Port Harcourt, Nigeria, in 2003 and 2007, respectively, where he was retained as a faulty member from 2005 until the present day. He has also been a visiting faculty member at the Abu Dhabi National Oil Company (ADNOC) Petroleum Institute. He was also a visiting faculty member at the Caledonian College of Engineering, Oman (Glasgow Caledonian University, UK). He is presently a visiting professor in the Department of Electrical and Computer Engineering, National University of Science and Technology (NUST), Oman, and an adjunct professor in the Department of Electrical and Electronic Engineering, Nisantasi University, Turkey. He was recognized as a top 1% peer reviewer in Engineering by Publons in 2018 and 2019 and was the editor's pick in the *Journal of Renewable and Sustainable Energy* in 2018. Dr. Okedu has published several books and journals/transactions in the field of renewable energy. He is an editor for including *Frontiers in Renewable Energy Research* (Smart Grids), *Energies* (MDPI), *International Journal of Smart Grids*, *International Journal of Electrical Engineering*, *Mathematical Problems in Engineering* and *Trends in Renewable Energy*. His research interests include power system stability, renewable energy systems, stabilization of wind farms, stability analysis of Doubly Fed Induction Generators (DFIGs) and Permanent Magnet Synchronous Generators (PMSG) variable speed wind turbines, augmentation and integration of renewable energy into power systems, grid frequency dynamics, wind energy penetration, FACTS devices and power electronics, renewable energy storage systems and hydrogen and fuel cells. Dr. Okedu was listed in the global Stanford University list of the top 2% of the world most academically cited researchers in the Scopus Worldwide Database. He also won the Outstanding Publication Award for publishing most Scopus-indexed papers in the year 2021-2022 at the National University.

1 Overview of Wind Energy Installations and Wind Turbine Technologies

1.1 OVERVIEW OF GLOBAL WIND ENERGY INSTALLATIONS

The global cumulative wind power capacity is around 743 Gigawatts (GW), as a result of the recent installation of 93 GW, in 2020 [1]. There was a 59% increment in the wind onshore market in 2020, accounting for 86.9 GW more installations, compared to 2019. The two countries taking the lead for onshore wind markets globally are China and the USA, with new onshore additions and increased market share ranging from 15 to 76%. Also, in the year 2020, on the regional level, onshore wind installations in Asia Pacific, North America and Latin America were on the rise. The total new wind installations in these three regions was 74 GW of new onshore wind capacity or 76% more than in 2019. As a result of the slow recovery of onshore installations in Germany in recent times, Europe experienced only 0.6% new onshore installations, while in Africa and the Middle East only 8.2 GW wind onshore installations were observed, with little or no considerable increment with the previous year.

Globally, in the offshore market, 6.1 GW was added in 2020, which is one of the highest wind offshore installations, with China taking the lead in installing half of the new offshore wind capacity. Europe experienced steady growth, with the Netherlands, Belgium, the UK, Germany and Portugal, taking leads, accordingly. Besides, in 2020, the United States and South Korea shared the remaining new offshore wind installations, making the cumulative offshore wind power capacity more than 35 GW or 4.8% of global cumulative wind power capacity [1]. Though, the global wind market growth would likely slow down in the near future because of an anticipated drop in onshore wind installations in China and in the United States as a result of incentive schemes that would expire [2, 3]. However, the wind market outlook still remains promising, with an expectation of over 469 GW of new wind onshore and offshore wind capacity in the next five years (accounting for about 94 GW of yearly new installations until 2025), considering present and new policies [1, 4].

Also, there was a milestone commitment to carbon neutrality in 2020, with the European Union, Japan, South Korea, Canada and South Africa each pledging to achieve net zero by the year 2050. In addition to China's net zero by 2060 target and the United States by 2050, the current net zero targets is 2/3 of the global economy, representing 63% of global emissions [5, 6]. Based on these facts, it is obvious that the era of fossil fuels is over and quickly taken over by the global energy transition.

DOI: 10.1201/9781003350910-1

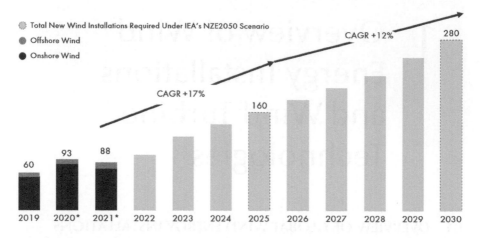

FIGURE 1.1   Global Annual Wind Installations to achieve Net Zero scenario by 2050.

Source: GWEC Market Intelligence; IEA World Energy Outlook (2020), volume in 2022–2024 and 2026–2029 are estimates.

There is no doubt that the wind industry has demonstrated incredible resilience, recently. The proliferation of wind farms and renewable energy is required to help limit global warming to below 2 °C, based on the Paris Agreement. Figure 1.1 shows the total new wind installations required by International Energy Agency (IEA), to achieve net zero by 2050 [7]. Figure 1.1 reflects that the annual wind installations must increase steadily in order to achieve net zero by 2050, by the addition of new global wind installations in (GW), ranging from 60 GW in 2019 to 280 GW in 2030, with an increased Compound Annual Growth Rate (CAGR) value of 17% at the end of 2025 and 12% at the end of 2030 [1, 7].

Figure 1.2 shows that in the year 2020, the new installations in the onshore wind market got up to 86.9 GW, while the offshore wind market was 6.1 GW, reflecting a considerable high onshore and offshore wind power installations around the globe. This was driven basically by China, Asia Pacific, where wind power continues to take the lead in global wind power developments, increasing its global market region by 8.5% in 2020 [1]. Another driving force for global wind power installation is the United States, and North America with a global market share of 18.4%, replacing Europe with 15.9%, as the second-largest regional market for new wind power installations. Latin America holds the fourth-largest regional market wind power installations with 5.0% in 2020, followed by Africa and the Middle East with regional wind power installations of 0.9%, in the same year. Figure 1.3 shows the new wind power capacity by region in 2020, while Figure 1.4 shows the global top five markets in the year 2020 for new wind power installations, where China, the United States, Brazil, Netherlands and Germany are taking the lead with a total of 80.6% of global wind power installations. Based on the cumulative wind power installations, as of the end of the year 2020, the top five markets are the same. The markets are China, the United States, Germany, India and Spain, accounting for 73% of global wind power installations.

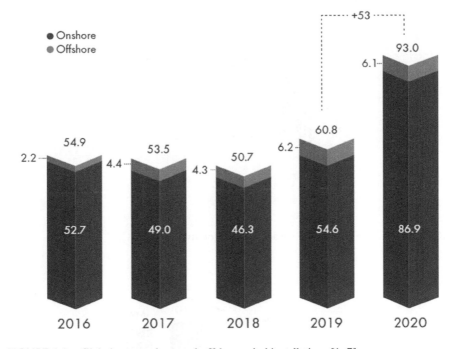

FIGURE 1.2 Global new onshore and offshore wind installations [1, 7].

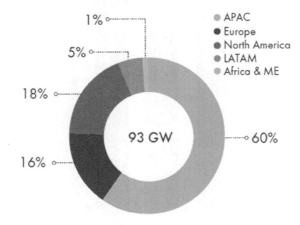

FIGURE 1.3 New wind power capacity installations by region [1].

The global offshore wind industry recorded over 6 GW of new installation in 2020, despite the effect of COVID-19. China took the lead in global new offshore wind installations for consecutive three years with over 3 GW of new offshore wind capacity recorded in 2020, as shown in Figure 1.4. In Europe, steady growth in offshore wind power installations was achieved, with the Netherlands taking the lead of nearly 1.5 GW new offshore wind power, then Belgium with 706 MW, the UK with 483 MW and Germany with 237 MW, respectively. Apart from China and Europe,

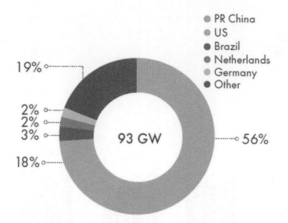

FIGURE 1.4   New wind power capacity for the top five markets [1].

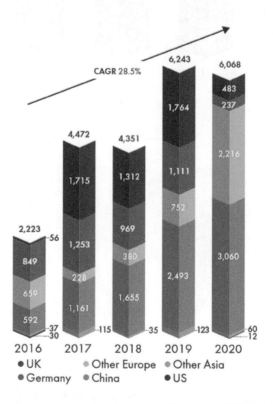

FIGURE 1.5   New global offshore wind power installations in 2020 [1, 7].

South Korea with 60 MW and the United States with 12 MW were the two other countries that recorded high new offshore wind installations in the year 2020, given in Figure 1.5 [7, 8]. More so, in the same year, in Portugal two new floating offshore wind turbines, were commissioned, making a total of 16.8 MW. A comparison of both onshore and offshore wind power capacity changes between 2019 and 2020 is given in Figure 1.6, where all regions increased installations except Europe, Africa and the Middle East. Figure 1.7 shows the major wind turbine manufacturers, where Vestas and General Electric are taking the lead.

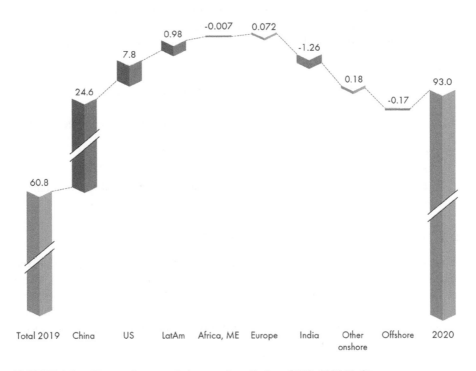

FIGURE 1.6 Changes in new wind power installations 2019–2020 [1, 2].

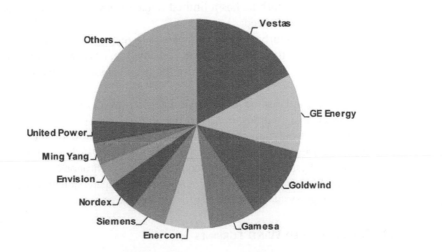

FIGURE 1.7 Wind turbine manufacturers.

1.2 CLASSIFICATION OF WIND ENERGY CONVERSION SYSTEM

Wind energy conversion systems can be classified as follows.

(1) They are two broad classifications based on the axis of the machine:
   (a) Horizontal Axis Machines: The axis of rotation is horizontal and the aero turbine plane is vertical, facing the wind.
   (b) Vertical Axis Machines: The axis of rotation is vertical. The sails or blades may also be vertical.
(2) They may be classified according to size:
   (a) Small-scale machines (up to 2 kW): Used in low-power applications.
   (b) Medium-size machines (2–100 kW)
   (c) Large-scale or large-size machines (100 kW and up): Used to generate power for distribution in central power grids.
(3) Classification as per type of output power
   (a) DC output: DC generators, Alternator rectifiers
   (b) AC output: Variable frequency variable or constant voltage, constant frequency variable or constant voltage

1.3 OVERVIEW OF WIND TURBINE TECHNOLOGIES

Wind turbine technologies could be classified into two groups as follows: Fixed Speed Wind Turbines (FSWTs) and Variable Speed Wind Turbines (VSWTs). There has been tremendous improvement in wind energy technology over the years, as a result of the fast innovations and developments in power electronics [9, 10]. This has resulted in the replacement of FSWTs with VSWTs. The following are some of the features of both wind turbine technologies.

1.3.1 FIXED SPEED WIND TURBINES

- This class of wind turbine has a limited range of power capture because it operates using fixed speed.
- This class of wind turbine lacks voltage and frequency control capability.
- This class of wind turbine is rugged in construction, has low running cost, is maintenance-free, with operational simplicity and possesses superior brushless features.
- This class of wind turbine needs large reactive power compensation during the transient state, in order to recover the air gap flux.
- This class of wind turbine system is expensive because of the installation of external reactive power compensation devices, such as Flexible AC Transmission Systems (FACTS) that could be Static Synchronous Compensators (STATCOM), Energy Capacitor Systems (ECS) or Superconducting Magnetic Energy Storage System (SMES), to provide reactive power.

1.3.2 VARIABLE SPEED WIND TURBINES

- This class of wind turbine has high energy conversion efficiency, during low and high winds because it has variable speed operation.

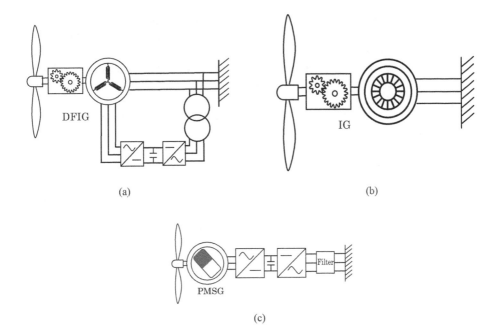

(a)

(b)

(c)

FIGURE 1.8 Wind turbine technologies. (a) Fixed speed induction generator wind turbine, (b) doubly fed induction generator variable speed wind turbine and (c) permanent magnet synchronous generator variable speed wind turbine.

- This class of wind turbine has less acoustic noise and mechanical stress.
- The class of wind turbine has better power quality in power grids without support from external reactive power compensators like the FACTS devices.
- This class of wind turbine employs power converters for secondary excitation, between 20 and 30% for a DFIG system and 100% for a PMSG system.
- This class of wind turbine has a lower cost of operation because it can generate power to the grid and the same time help in providing reactive power support to achieve stability of the network.

FSWTs are based on Squirrel Cage Induction Generator (SCIG), while VSWTs are based on Doubly Fed Induction Generator or Permanent Magnet Synchronous Generator (PMSG), as shown in Figure 1.8 (a–c) for FSWTs and VSWTs, respectively. A brief distinction between the three types of wind turbine-driven generators is given below.

1.4 OVERVIEW OF DFIG AND PMSG WIND TURBINES

Variable speed turbines are becoming the norm for new wind farm installations, because of high energy capture efficiency, coupled with reduced drive train stresses [11]. The PMSG VSWT also known as the direct-drive synchronous generator with permanent magnet excitation and the DFIG VSWT, with doubly fed back-to-back power converter type technologies, have become the two generator alternatives. The former has the disadvantage of cost mainly due to a fully rated power converter of 100% for energy capture. Although in the latter, a gearbox is needed, the DFIG requires a converter of only 20–30% of the generator's rating for an operating speed range of 0.7–1.3 per unit (p.u), resulting in a lower cost. This book would be focusing on the DFIG and PMSG VSWTs, since modern wind farms are built using both wind turbines.

Although the DFIG is not as rugged and robust as the squirrel-cage wind turbine type, its brushes have little wear and sparking when compared to DC machines and it is the only acceptable option for alternative energy conversion in the megawatts power range. With the help of modern power electronic devices, it is possible to recover the slip power dissipated in resistances [12]. The DFIG wind turbine uses a back-to-back power inverter system connected between the rotor side and the grid side of the machine, while the stator is directly connected to the grid. The DFIG can effectively operate at a wide range of speeds, based on the available wind speed and other specific operational requirements. Thus, it allows for better capture of wind energy [13, 14], and dynamic slip and pitch control may contribute to rebuilding the voltage at the wind turbine terminals and, at the same time, maintaining the power system stability after clearance of an external short-circuit fault [15]. Besides, DFIG wind turbine has shown better behavior regarding system stability during short-circuit faults in comparison to SCIG, because of its ability to decouple the control of active and reactive power output. The superior dynamic performance of the DFIG is achieved from the frequency or power converters which typically operate with sampling and switching frequencies of above 2 kHz [16]. At lower voltages down to 0%, the IGBTs (Insulated Gate Bipolar Transistors) of the DFIG are switched off and the system remains in standby mode [17–21]. If the voltages are above a certain cutoff or threshold value during grid disturbances, the DFIG wind turbine system is very quickly synchronized and is back in operation again.

The VSWT PMSG is connected through a back-to-back converter to the grid. The PMSG wind turbine is tied via the Grid Side Converter (GSC), control systems and Machine Side Converter (MSC) to the power network. This provides maximum flexibility, enabling full real and reactive power control and fault ride through capability during voltage dips, as compared to the VSWT DFIG technology. However, the use of this wind turbine technology is limited when compared to the DFIG technology due to high cost. Compared to the widely used DFIG wind turbine, the PMSG-based VSWT has a more feasible technology that is promising in wind generation because it is self-excited, hence, is possible to operate with higher efficiency and power factor. Besides, there is no gearbox system because of its low rotational speed. Therefore, no careful and regular maintenance is required in this wind turbine topology, unlike the DFIG-based wind turbine [22]. In addition, the converters have room for flexible control of active and reactive dissipation of power in normal and transient states [23, 24]. However, some drawbacks of the PMSG wind turbine are complex construction and controller control strategy, compared to the traditional FSWT based on SCIGs.

1.5 DFIG WIND TURBINE MODELING

A wind turbine is an electromechanical energy conversion device that captures kinetic energy from the wind and turns it into electrical energy. The primary components of a wind turbine for modeling purposes consist of the turbine rotor or prime mover, a shaft and a gearbox unit (a speed changer) [25]. The dynamics interaction involving forces from the wind and the response of wind turbine determines the amount of kinetic energy that can be extracted. The aerodynamic torque and the mechanical power of a wind turbine are given as follows [26–28].

$$T_{\mathrm{M}} = \frac{\pi \rho R^3}{2} V_{\mathrm{w}}^2 C_{\mathrm{t}}(\lambda)[\mathrm{NM}] \tag{1.1}$$

$$P_{\mathrm{M}} = \frac{\pi \rho R^2}{2} V_{\mathrm{w}}^3 C_{\mathrm{p}}(\lambda)[\mathrm{W}] \tag{1.2}$$

where ρ is the air density, R is the radius of the turbine, V_{w} is the wind speed, $C_{\mathrm{p}}(\lambda,\beta)$ is the power coefficient given by

$$C_{\mathrm{p}}(\lambda,\beta) = 0.5(\Gamma - 0.02\beta^2 - 5.6)e^{-0.17\Gamma} \tag{1.3}$$

The relationship between C_{t} and C_{p} is

$$C_{\mathrm{t}}(\lambda) = \frac{C_{\mathrm{p}}(\lambda)}{\lambda} \tag{1.4}$$

$$\lambda = \frac{\omega_{\mathrm{r}} R}{V_{\mathrm{w}}} \tag{1.5}$$

In (1.3), $\Gamma = \dfrac{R}{\lambda} \dfrac{(3600)}{(1609)}$ and in (1.5), λ is the tip speed ratio.

The wind turbine characteristics [29] for both IGs and DFIGs are shown in Figures 1.9 and 1.10, respectively. In Figure 1.10, the power capture characteristics of the

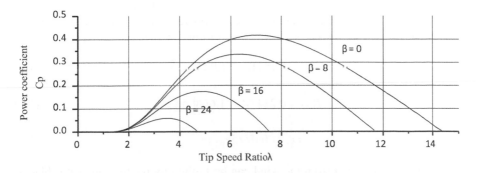

FIGURE 1.9 CP- λ curves for different pitch angles (for FSWT).

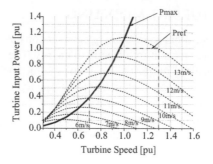

FIGURE 1.10 Turbine characteristic with maximum power point tracking (for DFIG VSWT).

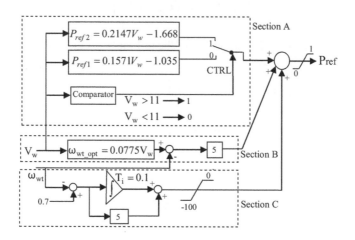

FIGURE 1.11 Control block to determine active power reference P_{ref}.

turbine with respect to the rotor speed are shown. The dotted lines show the locus of the maximum power point of the turbine, which is used to determine the reference of active power output P_{ref}. Equations (1.6) and (1.7) are used to calculate the reference of the active power output P_{ref}, as shown in Section A of Figure 1.11. The optimum rotor speed ω_{ropt} is given in Equation (1.8). The operation range of the rotor of DFIG is chosen from 0.7 to 1.3 pu, as shown in the turbine characteristics in Figure 1.10.

$$P_{\text{ref1}} = 0.1571V_{\text{w}} - 1.035 \left[\text{pu}\right] \tag{1.6}$$

$$P_{\text{ref2}} = 0.2147V_{\text{w}} - 1.668 \left[\text{pu}\right] \tag{1.7}$$

$$\omega_{\text{opt}} = 0.0775V_{\text{w}} \left[\text{pu}\right] \tag{1.8}$$

The power extracted from the wind can be limited by pitching the rotor blades. The angle control is usually done with a Proportional Integral (PI) controller in such

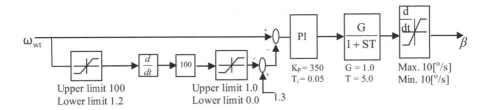

FIGURE 1.12 Pitch controller for DFIG VSWT.

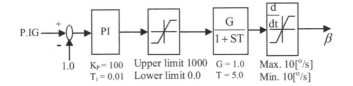

FIGURE 1.13 Pitch controller for SCIG FSWT.

a way that the pitch controller shown in Figure 1.12 controls the angle when the rotor speed exceeds 1.3 pu for the case of DFIG that operates in variable speed mode. Figure 1.13 shows the pitch controller for the fixed speed Wind Turbine Generator System (WTGS). In order to get a realistic response in the pitch angle control system, the servomechanism accounts for a servo time constant, T_{servo} and a limitation of both the pitch angle and its rate of change, as shown in Figures 1.12 and 1.13, respectively. The rate of change limitation is very important during grid fault, because it decides how fast the aerodynamic power can be reduced in order to prevent over-speeding during fault [30, 31]. Considering the realistic scenario for a heavy mechanical system, the rate limiter must be incorporated to simulate the pitch controller. Therefore the pitch rate limiter of ±10 deg./sec. is used for both pitch controllers in this text.

1.6 PMSG WIND TURBINE MODELING

The PMSG turbine is made up of a generator blade, system controller, components of power electronics and transformer [32]. The wind turbine is tied to the power grid via its back-to-back full-power converters that are capable of converting wind into electrical energy. To realize wind energy maximum power tracking [33, 34], the motor speed or torque of the wind generator is controlled by the MSC. The stabilization of the voltage of the DC-link and regulation of the power factor and quality of the wind generator is done basically by the GSC.

The mechanical power extracted by the wind generator can be expressed as [35]:

$$P_w = \frac{1}{2}\rho\pi R^2 V_w^3 C_p\left(\lambda,\beta\right) \tag{1.9}$$

From Equation (1.9), P_w is the wind power that is captured, expressed in (W), the air density is ρ, expressed in (kg / m^3), the radius R is expressed in (m) and the wind speed V_w is expressed in (m/s). The wind generator's power coefficient is C_p and is related to the ratio of the tip speed (λ) and angle of the pitch (β), respectively, as expressed in Equation (1.91) [36].

$$C_p(\lambda, \beta) = c_1 \left(\frac{c_2}{\lambda_i} - c_3\beta - c_4 \right) e^{\frac{-c_5}{\lambda_i}} + c_6\lambda \tag{1.10}$$

where

$$\frac{1}{\lambda_i} = \frac{1}{\lambda - 0.08\beta} - \frac{0.035}{\beta^3 + 1} \tag{1.11}$$

In Equation (1.10), c_1 to c_6 are the characteristic coefficients of the wind turbine. In the PMSG wind turbine, the Maximum Power Point Tracking (MPPT) is based on the rotor speed and the maximum power could be obtained by [37]:

$$P_{\text{MPPT}} = \frac{1}{2} \rho \pi R^2 \left(\frac{\omega_r R}{\lambda_{\text{opt}}} \right)^3 c_{\text{popt}} \tag{1.12}$$

where λ_{opt} is the optimal value of λ, c_{popt} is the optimal power coefficient and ω_r is the rotor speed of the wind generator. The wind generator characteristics relating to the turbine output power and the rotor speed for varying wind speeds are shown in Figure 1.14. The maximum obtainable power output is 1.0 pu at 12 m/s occurring at

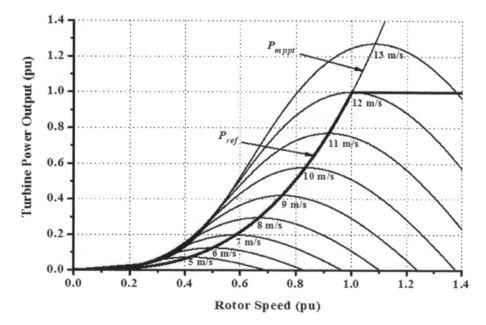

FIGURE 1.14 Turbine characteristic with maximum power point tracking (for PMSG VSWT).

1.0 pu rotational speed. It should be noted that the reference power P_{ref} of the wind turbine is limited to the wind generator rated power.

The rotating frame based on d-q reference for the dynamic model in the PMSG wind turbine is expressed as [38]:

$$\frac{d\psi_{sd}}{dt} = -V_{sd} - R_s I_{sd} - \omega_e \psi_{sq} \tag{1.13}$$

$$\frac{d\psi_{sq}}{dt} = -V_{sq} - R_s I_{sq} - \omega_e \psi_{sd} \tag{1.14}$$

From Equations (1.13) and (1.14),

$$\psi_{sd} = \left(L_{sd} + L_{md}\right) I_{sd} + \psi_m \tag{1.15}$$

$$\psi_{sq} = \left(L_{sq} + L_{mq}\right) I_{sq} \tag{1.16}$$

where V_{sd} and V_{sq} are the voltages of the stator circuit, R_s is the winding resistance of the stator, I_{sd} and I_{sq} are the currents in the stator d and q reference frames, ω_e is the rotational speed of the wind generator, ψ_{sd} and ψ_{sq} are the flux linkages of the stator circuit, L_{sd} and L_{sq} are the stator wind leakage inductances, L_{md} and L_{mq} are the magnetizing inductances and ψ_m is the linkage flux of the machine's permanent magnet.

Putting Equations (1.15), (1.16) into (1.13), (1.14), the PMSG differential equations could be obtained as:

$$L_d \frac{dI_{sd}}{dt} = -V_{sd} - R_s I_{sd} - \omega_e L_q I_{sq} \tag{1.17}$$

$$L_q \frac{dI_{sq}}{dt} = -V_{sq} - R_s I_{sq} + \omega_e L_d I_{sd} + \omega_e \psi_m \tag{1.18}$$

With

$$L_d = L_{sd} + L_{md} \tag{1.19}$$

$$L_q = L_{sq} + L_{mq} \tag{1.20}$$

The PMSG active and reactive powers are given as:

$$P_s = V_{sd} I_{sd} + V_{sq} I_{sq} \tag{1.21}$$

$$Q_s = V_{sq} I_{sd} - V_{sd} I_{sq} \tag{1.22}$$

The wind generator electrical torque (T_e) with number of pole pairs is given as:

$$T_e = 0.5 p \left(\psi_m I_{sq} + \left(L_d - L_q\right) I_{sd} I_{sq}\right) \tag{1.23}$$

1.7   CHAPTER CONCLUSION

Recently, the penetration of wind energy into power grids is on the rise globally. This is evident from the wind energy installation statistics presented in this chapter, considering the various regions in the world and the top markets for wind energy. China and Asia Pacific continue to take the lead in global wind power developments, increasing its global market region by 8.5% in recent years. The USA and North America have a global market share of 18.4% and have recently replaced Europe with 15.9% as the second-largest regional market for new wind power installations. Latin America with 5.0% and Africa and the Middle East region with 0.9% hold the fourth and fifth positions in global wind installations, respectively. China, the USA, Brazil, the Netherlands and Germany have 80.6% of global wind power installations. The recent top five countries in wind power installations are China, the USA, Germany, India and Spain, possessing 73% of global wind power installations.

The VSWT technology is superior to the fixed speed with turbine technology in wind energy conversion system due to their high wind energy capture efficiency. The DFIG and the PMSG are the two main types of VSWTs employed in modern wind farms. This chapter presented the basic modeling of both wind turbines considering their MPPT characteristics.

REFERENCES

[1] Global Wind Energy Council (GWEC), *Annual Report*, 2021.

[2] https://unfccc.int/climate-action/race-to-zero-campaign;   https://eciu.net/analysis/briefings/net-zero/net-zero-thescorecard#:~:text=Net%20zero%20economies, (World%20Bank%2C%202018)

[3] https://www.iea.org/reports/world-energy-outlook-2020

[4] https://www.carbonbrief.org/analysis-world-has-already-passed-peak-oil-bp-figures-reveal.

[5] https://www.tradewindsnews.com/offshore/-51bn-in-wind-farm-capital-spending-outstrips-oil-and-gas-for-firsttime/2-1-955552

[6] http://www3.weforum.org/docs/WEF_Net_Zero_Challenge_The_Supply_Chain_Opportunity_2021.pdf; https://climateactiontracker.org/publications/global-update-paris-agreement-turning-point/

[7] International Energy Agency (IEA), *Annual Report*, 2020.

[8] World Energy Outlook, International Renewable Agency (IRENA), Global renewables outlook: Energy transformation 2050, 2020, https://www.irena.org/publications/2020/Apr/Global-Renewables-Outlook-2020

[9] K. E. Okedu, S. M. Muyeen, R. Takahashi, and J. Tamura, "Protection schemes for DFIG considering rotor current and DC-link voltage," *24th IEEE-ICEMS (International Conference on Electrical Machines and System)*, Beijing, China, August 2011, pp. 1–6.

[10] MNES, "Ministry of non-conventional energy sources," *Annual Report*, 2005.

[11] K. E. Okedu, "Introductory," In *Power System Stability*, Kenneth E. Okedu (Ed.), United Kingdom: INTECH, 2019, pp. 1–10.

[12] S. Bozhko, G. Asher, R. Li, J. Clare, and L. Yao, "Large offshore DFIG-based wind farm with line-commutated HVDC connection to the main grid: Engineering studies," *IEEE Transactions on Energy Conversion*, vol. 23, no. 1, pp. 119–127, 2008.

[13] M. Godoy Simoes, and F. A. Farret, *Renewable Energy Systems: Design and Analysis with Induction Generators*, Boca Raton, FL: CRC Press, LLC, 2004.

[14] S. Santos, and H. T. Le, "Fundamental time-domain wind turbine models for wind power studies," *Renewable Energy*, vol. 32, pp. 2436–2452, 2007.

[15] M. Haberberger, and F. W. Fuchs, "Novel protection strategy for current interruptions in IGBT current source inverters," *Proceedings EPE-PEMC*, Oslo, Norway, 2004.

[16] K. E. Okedu, S. M. Muyeen, R. Takahashi, and J. Tamura, "Wind farms fault ride through using DFIG with new protection scheme," *IEEE Transactions on Sustainable Energy*, vol. 3, no. 2, pp. 242–254, 2012.

[17] A. A. El-Sattar, N. H. Saad, and M. Z. S. El-Dein, "Dynamic response of doubly fed induction generator variable speed wind turbine under fault," *Electric Power System Research*, vol. 78, pp. 1240–1246, 2008.

[18] T. Takahashi, "IGBT protection in AC or BLDC motor drives," *Technical Paper, International Rectifier*, El Segundo, CA, 2004.

[19] II. Xie, "Voltage source converters with energy storage capability," Ph.D Thesis, Royal Institute of Technology, School of Electrical Engineering, Division of Electrical Machines and Power Electronic, Stockholm, Sweden, 2006.

[20] B. H. Chowdhury, and S. Chellapilia, "Doubly-fed induction generator control for variable speed wind power generation," *Electric Power System Research*, vol. 76, pp. 786–800, 2006.

[21] H. Karim-Davijani, A. Sheikjoleslami, H. Livani, and M. Karimi-Davijani, "Fuzzy logic control of doubly fed induction generator wind turbine," *World Applied Science Journal*, vol. 6, no.4, pp. 499–508, 2009.

[22] M. Ammar, and M. E. Ammar, "Enhanced flicker mitigation in DFIG based distributed generation of wind power," *IEEE Transactions on Industrial Informatics*, vol. 12, no. 6, pp. 2041–2049, 2016.

[23] M. Rosyadi, S. M. Muyeen, R. Takahashi, and J. Tamura. "A design fuzzy logic controller for a permanent magnet wind generator to enhance the dynamic stability of wind farms," *Applied Sciences*, vol. 2, pp. 780–800, 2012, doi: 10.3390/app2040780

[24] G. Michalke, A. D. Hansen, and T. Hartkopf, "Control strategy of a variable speed wind turbine with multi-pole permanent magnet synchronous generator," *Proceedings of European Wind Energy Conference and Exhibition*, Milan, Italy, 7–10 May 2007.

[25] K. E. Okedu, *Enhanced Power Grid Stability Using Doubly-Fed Induction Generator, American Institute of Physics Publishers (AIPP)*, New York: AIP Publishing LLC AIP Publishing Melville, 2020. doi: 10.1063/9780735422292

[26] K. E. Okedu, Onshore Wind *Farms: Dynamic Stability and Applications in Hydrogen Production, American Institute of Physics Publishers (AIPP)*, New York: AIP Publishing LLC AIP Publishing Melville, 2021. doi: 10.1063/9780735422995

[27] K. E. Okedu, R. Uhunmwangho, P. O. Madifie, and C. C. Chiodule, "Technical review of wind farm improved performance and environmental development challenges," In *Wind Farms: Performance, Economic Factors and Effects on the Environment*, M. Dunn (Ed.), New York: NOVA, 2016, pp. 1–47. ISBN: 978-1-634848411

[28] R. Takahashi, J. Tamura, M. Futami, M. Kimura, and K. Idle, "A new control method for wind energy conversion system using double fed synchronous generators," *IEEJ Transactions on Power and Energy*, vol. 126, no. 2, pp. 225–235, 2006.

[29] K. E. Okedu, S. M. Muyeen, R. Takahashi, and J. Tamura, "Wind farm stabilization by using DFIG with current controlled voltage source converters taking grid codes into consideration," *IEEJ Transactions on Power and Energy*, vol. 132, no. 3. pp. 251–259, 2012.

[30] M. Garcia-Garcia, M. P. Comech, J. Sallan, and A. Liombart, "Modelling wind farms for grid disturbances studies," *Science Direct, Renewable Energy*, vol. 33, pp. 2019–2121, 2008.

[31] R. Babouri, D. Aouzellag, and K. Ghedamsi, "Integration of doubly fed induction generator entirely interfaced with network in a wind energy conversion system," *Terra Green 13 International Conference-Advancements in Renewable Energy and Clean Environment, Energy Procedia Science Direct*, vol. 36, pp. 169–178, 2013.

[32] Y. Li, Z. Xu, and K. P. Wong, "Advanced control strategies of PMSG-based wind turbines for system inertia support," *IEEE Transactions on Power Systems*, vol. 32, pp. 3027–3037, 2017.

[33] N. Priyadarshi, V. Ramachandaramurthy, S. Padmanaban, and F. Azam, "An ant colony optimized MPPT for standalone hybrid PV-wind power system with single Cuk converter," *Energies*, vol. 12, pp. 167, 2019.

[34] S. W. Lee, and K. H. Chun, "Adaptive sliding mode control for PMSG wind turbine systems," *Energies*, vol. 12, pp. 595, 2019.

[35] S. Heier, "Wind energy conversion systems," In *Grid Integration of Wind Energy Conversion Systems*, Chicester, UK: John Wiley & Sons Ltd, 1998, pp. 34–36.

[36] MathWorks, MATLAB documentation center. http://www.mathworks.co.jp/jp/help/ (accessed on 12 March 2012).

[37] S. M. Muyeen, A. Al-Durra, and J. Tamura, "Variable speed wind turbine generator system with current controlled voltage source inverter," *Energy Conversion and Management*, vol. 52, no. 7, pp. 2688–2694, 2011.

[38] S. Li, T. A. Haskew, and L. Xu, "Conventional and novel control design for direct driven PMSG wind turbines," *Electric Power Systems Research*, vol. 80, pp. 328–338, 2010.

2 DFIG with Different Inverter Schemes

2.1 CHAPTER INTRODUCTION

In wind energy applications, the Doubly Fed Induction Generator (DFIG) has one main merit of utilizing only 20–30% of the wind generator rating for the power converters linking the rotor side and the grid side [1, 2]. Lately, the use of Insulated Gate Bipolar Transistors (IGBTs) is on the rise for high-power applications among semiconductor devices. This is because the current capability of the IGBT switches can be increased by configuring them in parallel connection [3]. Because the IGBT switches have high power and are required to be subjected to high voltage and current, their transient operations during periods of several microseconds are vital [4, 5]. Therefore, the ability for variable speed wind turbine IGBT based on withstanding abnormal conditions is strictly paramount to achieve the lifetime specifications [6, 7] of the wind turbine and at the same time to fulfill the requirement of grid codes.

As the use of multilevel converters is becoming popular in wind energy conversion systems, because of their robustness during transient conditions, this chapter tends to improve the performance of the six-step 2-level IGBT inverter by proposing a coordinated hybrid control of a new Phase Lock Loop (PLL) configuration and a Series Dynamic Braking Resistor (SDBR). The high voltage usually experienced during grid disturbances is shared by the small inserted resistance because of the series connection strategy of the braking resistor employed in this study. Thus, the loss of the converter control system is not experienced in this topology due to the effects of induced overvoltage. In addition, the series-connected braking resistor strategy significantly reduces the very high current in the rotor circuitry of the wind generator to lower values. Consequently, the overvoltage of the DC-link that was supposed to be dangerous to the power converters of the wind generator is avoided because of the low DC-link capacitor charging current [8–10]. Although many studies in the literature considered the use of fault current limiter for enhancing power quality and limiting fault current of DFIG wind turbines in wind farms [11–13], in this chapter, the preferred position of the braking resistor in the wind generator system was analyzed considering different switching signals. The optimal braking resistor position and switching strategy were used for further analysis of the proposed wind generator schemes used in this study. In addition, a comparative study using the proposed scheme for the 2-level IGBT inverter was carried out with the schemes having parallel interleaved IGBT inverter and 3-level IGBT inverter. Simulations were run in PSCAD/EMTDC [14]. The proposed hybrid scheme could help to increase the current capability and post-fault recovery of the wind turbine. In addition, the space vector modulation of the inverter schemes resulted in maximum value change in Common Mode (CM) voltage, using the proposed hybrid control strategy of the PLL

DOI: 10.1201/9781003350910-2

and SDBR scheme [15]. Consequently, there is improved switched output voltage of the converter leg of the voltage source converter. The results show that the hybrid scheme in the various inverter topologies considered in the study can enhance the performance of the wind generator variables during severe three-phase-to-grid fault.

2.2   MODEL SYSTEM OF STUDY

The model system of study is shown in Figure 2.1. The DFIG wind turbine modeling can be referred to in Chapter 1 of this book. In Figure 2.1, the rotor side (A), the grid side (B) and the stator side (C) of the wind generator show the insertion of the proposed braking resistor scheme in a series connection, respectively. A three-phase-to-ground fault was used to investigate the optimal position of the braking resistor considering the model system of study in Figure 2.1, which is connected to an infinite bus. Figure 2.2 shows the various control signals, DC-link voltage, current of the rotor for the wind generator and terminal voltage of the grid, along with the conditions of operation of the braking resistor. The optimal position and switching strategy of the braking resistor in the wind generator were used for further analysis of the inverter schemes proposed in this chapter.

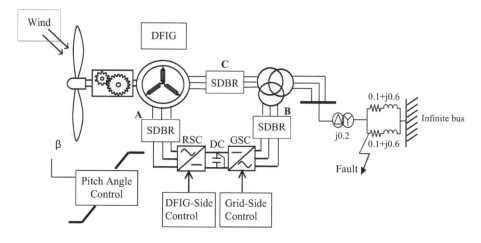

FIGURE 2.1   Model system.

$$E_{dc} < 1.5E_{dc} \text{ or } I_r < 2I_r \text{ or } V_g > 0.9: 1 \text{ Normal condition}$$

$$E_{dc} > 1.5E_{dc} \text{ or } I_r > 2I_r \text{ or } V_g < 0.9: 0 \text{ Fault condition}$$

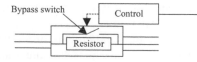

FIGURE 2.2   SDBR control strategy.

2.3   VARIABLE SPEED DRIVE CONTROL

Figure 2.3 shows the control strategy of the DFIG-based variable speed wind turbine. The cost of the crowbar protection scheme used in a conventional DFIG system is more than the other schemes like the braking resistor or DC chopper. During grid fault, the crowbar makes the variable speed wind generator based on DFIG act like a fixed speed wind generator. This is done by disconnecting the wind generator's rotor side converter. In this chapter, the DC-link (chopper) scheme is the alternative employed in place of the traditional crowbar switch, as shown in Figure 2.3. The coordinated controls of the active and reactive power of the DFIG system via abc to dq and dq to abc conversion to generate pulses for the switching of the PWM are shown in Figure 2.3.

The new PLL control strategy used in this study has a delay element incorporated to improve the performance of the DFIG system. The PLL scheme is designed based

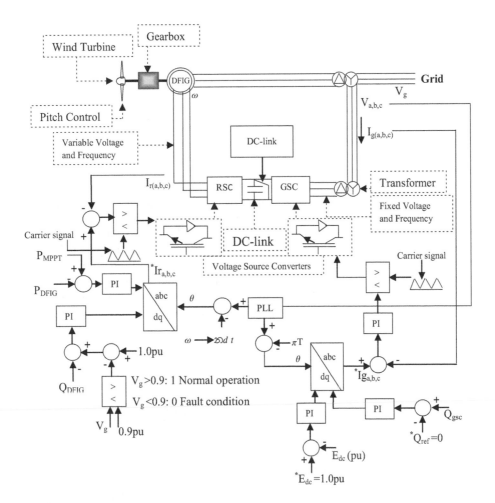

FIGURE 2.3   DFIG control strategy.

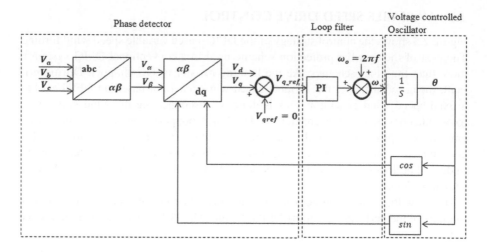

FIGURE 2.4   Conventional PLL scheme.

on a frequency of 50 Hz and is rated line-to-line voltage of 0.69 kV. Unlike the conventional PLL scheme, the proposed PLL scheme in this chapter integrates the Sine and Cosine function angles for the three phases via a multiplier before the insertion of a delay element, in order to boost its synchronizing strength with the grid for better performance during transient state.

Figure 2.4 shows the conventional three-phase PLL scheme, which is basically an error signal feedback system based on the principle of a synchronous rotating frame, with low pass filters and voltage-controlled oscillator. The working principle is based on the conversion of the measured voltage of a three-phase system to d-q component via conversion coordinate and set DC voltage reference of q-axis v_{ref}. Figure 2.5 shows the vector partition diagram for the PLL. From Figure 2.5, the d-axis component is fully co-phased with the vector voltage when the q-axis component is zero, despite the values of the voltages on the various q-axes. The Proportional Integral (PI) controller in Figure 2.4 helps in obtaining the frequency of the system. With the grid voltage having only positive sequence fundamental components, the d-q steady value coordinate is DC current and its phase and frequency can be locked by

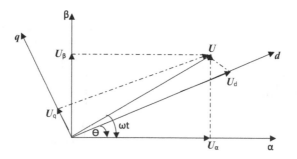

FIGURE 2.5   PLL vector partition.

controlling the q-axis component to zero. Frequency detection when the grid voltage is balanced is achieved for the conventional three-phase PLL based on tracking the grid voltage positive sequence fundamental components, as the inner of the PLL is closed loop controller. However, during the transient state, there will be a sudden change giving rise to instantaneous negative sequence and zero sequence fundamental components, which leads to the oscillation of the PLL output angle.

In the course of the grid voltage being unbalanced, there exist positive, negative and zero sequence fundamental components. For a typical three-phase system without a neutral point, the zero sequence is not usually considered. Thus, the grid voltage can be expressed as [16]:

$$v_{abc} = v_{abc}^+ + v_{abc}^- = V^+ \begin{vmatrix} \cos \omega t \\ \cos\left(\omega t - \frac{2}{3}\pi\right) \\ \cos\left(\omega t + \frac{2}{3}\pi\right) \end{vmatrix} + V^- \begin{bmatrix} \cos\left(\omega t + \theta_0^-\right) \\ \cos\left(\omega t + \frac{2}{3}\pi + \theta_0^-\right) \\ \cos\left(\omega t - \frac{2}{3}\pi + \theta_0^-\right) \end{bmatrix} \qquad (2.1)$$

From Equation (2.1), V^+, V^- gives the voltage amplitude separately for the positive and negative sequences, respectively. θ_0^- gives the relative phase angle of the initial voltage negative sequence. The voltage of the output is achieved after a 3/2 conversion in $\alpha\beta$ static coordinates and expressed as [17, 18]:

$$v_{\alpha\beta} = \begin{bmatrix} v_\alpha \\ v_\beta \end{bmatrix} = T_{\alpha\beta} v_{abc} \qquad (2.2)$$

$$T_{\alpha\beta} = \frac{2}{3} \begin{bmatrix} 1 & -\frac{1}{2} & -\frac{1}{2} \\ 0 & \frac{\sqrt{3}}{2} & -\frac{\sqrt{3}}{2} \end{bmatrix} \qquad (2.3)$$

In the static $\alpha\beta$ static coordinates of the grid voltage, the fundamental positive and negative sequence component is given as:

$$v_{\alpha\beta} = v_{\alpha\beta}^+ + v_{\alpha\beta}^- = V^+ \begin{bmatrix} \cos wt \\ \sin wt \end{bmatrix} + V^- \begin{bmatrix} \cos(wt + \theta_o^-) \\ -\sin(wt + \theta_o^-) \end{bmatrix} \qquad (2.4)$$

After d-q transformation to the synchronous coordinate system, the following equation is obtained:

$$v_{dq}^+ = T_{dq}^+ v_{\alpha\beta} = T_{dq}^+ v_{\alpha\beta}^+ + T_{dq}^+ v_{\alpha\beta}^- \qquad (2.5)$$

$$T_{dq}^+ = \begin{bmatrix} \cos\theta & \sin\theta \\ -\sin\theta & \cos\theta \end{bmatrix} \qquad (2.6)$$

Considering Equations (2.5) and (2.6) lead to

$$v_{dq}^+ = V^+ \begin{bmatrix} \cos wt - \theta \\ \sin wt - \theta \end{bmatrix} + V^- \begin{bmatrix} \cos(wt + \theta_o^- + \theta) \\ -\sin(wt + \theta_o^- + \theta) \end{bmatrix} \tag{2.7}$$

The effective working of the PLL requires that $\omega t \approx \theta$, so Equation (2.7) can be expressed as

$$\begin{cases} v_d^+ = V^+ + V^- \cos\left(2\omega t + \theta_0^-\right) \\ v_q^+ = V^- \sin\left(2\omega t + \theta_0^-\right) \end{cases} \tag{2.8}$$

From Equation (2.8), it could be observed that synchronous sequence due to the positive system coordinates enables the conversion to DC components from the positive sequence, and for the components of the negative sequence, a value that is twice the frequency component is obtainable. This is because the traditional PI controller can only be used to remove the steady state error, thus the negative voltage will have an influence on the output of the PLL.

The traditional PLL strategy has a negative sequence component during transient conditions in the grid. Therefore, a low loop filter cutoff frequency is required to help achieve improved performance of the system during steady state. Based on this, since the transient performance of the system would also be affected, this study aims to propose a new configuration and control topology of PLL with a delay element, as shown in Figure 2.6, in order to overcome this drawback. The negative effects due to the component of the voltage sequence would be greatly mitigated during transient conditions in the grid with the help of the delay element e^{-sT}.

The principle of the delay element is based on phase shifting after abc/dq transformation in controller, as shown in Figure 2.6, compared to the traditional strategy employed for the PLL in Figure 2.4. The proposed control PLL scheme would counteract twice the grid frequency disturbances based on Equation (2.9).

$$\begin{cases} v_d^+ = \cos 2\omega t + \cos 2\omega\left(t - \dfrac{\pi}{4}\right) \\ v_q^+ = \sin 2\omega t + \sin 2\omega\left(t - \dfrac{\pi}{4}\right) \end{cases} \tag{2.9}$$

2.4 DFIG NEUTRAL POINT CLAMPED MULTILEVEL CONVERTER TOPOLOGY

Figure 2.7 shows the model system for the DFIG wind generator NPC MLC topology. The NPC has three legs, A, B and C with three different voltage states. Switches 1 and 3 are complementary on each leg, therefore, when switch 1 is on, switch 3 is off and the other way round. Similarly switches 2 and 4 are complementary. From Figure 2.7, each of the capacitors has a constant voltage of $0.5V_{dc}$, therefore having

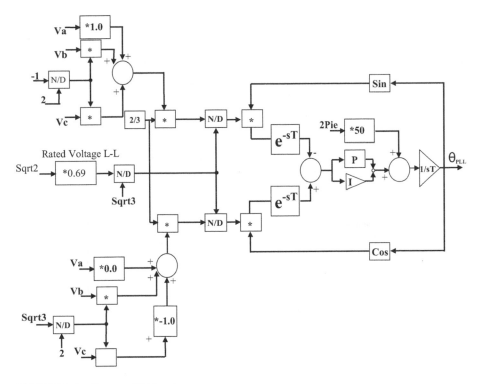

FIGURE 2.6 Proposed PLL scheme.

the two upper switches on will lead to an output voltage of V_{dc}, compared to 0 level. Also, when switches 2 and 3 are on, it would lead to $0.5V_{dc}$ and having the two lower switches on leads to an output voltage level of 0. There exists a forbidden state whose scenario is when the first switch is on and the second is off, in addition to the three states. Table 2.1 shows the bridge leg voltages at different combinations of the switch states. The excitation parameters of the DC circuitry and the DFIG wind generator parameters are given in [19].

2.5 DFIG PARALLEL INTERLEAVED MULTILEVEL CONVERTER TOPOLOGY

Figure 2.8 shows the model system for the DFIG wind generator Parallel Interleaved MLC topology [20]. Figure 2.9 shows the table of switching, with the switching sequence of the pulse width modulation represented by the numbers. The space vector reference V_{ref} is formulated based on the nearest three voltage vector summation geometry in region 1, as shown in Figure 2.10. The connections of the passive filters of the grid side to the parallel interleaved MLC are shown in the model system. The reduction of the harmonics is achieved by the grid filters of value 9.6 mH and also with the help of the CM inductor. The IGBTs are numbered for both RSC and GSC power converters.

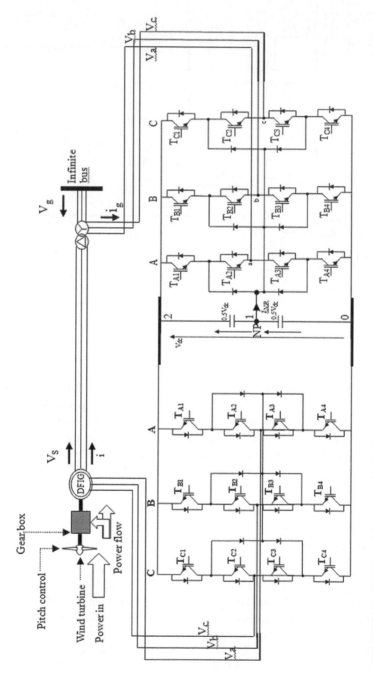

FIGURE 2.7   DFIG model system with NPC MLC scheme.

TABLE 2.1
DFIG NPC MLC Bridge Leg Voltages for Different Combinations of Switch States

State of leg	V_{a0}	T_{A1}	T_{A2}	T_{A3}	T_{A4}
2	V_{dc}	On	On	Off	Off
1	$0.5V_{dc}$	Off	On	On	Off
0	0	Off	Off	On	On

For the operation of the two-level conventional space vector modulation, the converters cycles through a switching cycle with four switch states. However, for the 180 degrees parallel interleaved MLC scheme, its space vector modulation gives voltage and the CM changes with a peak value to $\pm V_{dc}$, whereas for the clamped Discontinuous Pulse width Modulation (DPWM) schemes, this value is $\pm \dfrac{2V_{dc}}{3}$. The common leg flux can be mitigated by the time integral reduction of the CM voltage change. Avoiding the zero vector voltage would help achieve this purpose. In this chapter, the Near State PWM scheme having three nearest active voltage vectors to synthesize the reference voltage vector $\vec{V}_{ref}$ is used. The table of switching sequence in Figure 2.9, for the parallel interleaved MLC is formulated by the vector space diagram in Figure 2.10, with six regions, like the NPC MLC scheme earlier discussed. The formulation considers the voltage vectors corresponding to the sequence numbers. However, in the parallel interleaved MLC scheme, the 612 sequence of switching is employed in sub-sector 1, for $0° \leq \varphi \leq 30°$ and sub-sector 12, for $330° \leq \varphi \leq 360°$. Region 1 is formulated by the geometry for the voltage of reference $\vec{V}_{ref}$, as shown in Figure 2.10. Basically, the two sub-sections and the reference voltage make up region 1. The table of switching and their corresponding dwell times, based on $\vec{V}_1$, $\vec{V}_2$ and $\vec{V}_6$ active voltages are [3]:

$$T_1 = \left(\sqrt{3}\,\frac{V_{\alpha,r}}{V_{dc}} + \frac{V_{\beta,r}}{V_{dc}} - 1 \right) T_s \tag{2.10}$$

$$T_2 = \left(1 - \frac{2}{\sqrt{3}}\,\frac{V_{\alpha,r}}{V_{dc}} \right) T_s \tag{2.11}$$

$$T_6 = \left(1 - \frac{1}{\sqrt{3}}\,\frac{V_{\alpha,r}}{V_{dc}} - \frac{V_{\beta,r}}{V_{dc}} \right) T_s \tag{2.12}$$

T_1, T_2, and T_3, in Equations (2.10)–(2.12) are the active vector voltages dwell times, while the time of switching is T_s. The α, β components in Figure 2.10 are given by V_α, V_β, while the α and β components for region 1, at the beginning of the region are $V_{\alpha,r}$ and $V_{\beta,r}$. These are expressed mathematically as:

$$V_{\alpha,r1} = \frac{\sqrt{3}}{2} V_\alpha - \frac{1}{2} V_\beta \tag{2.13}$$

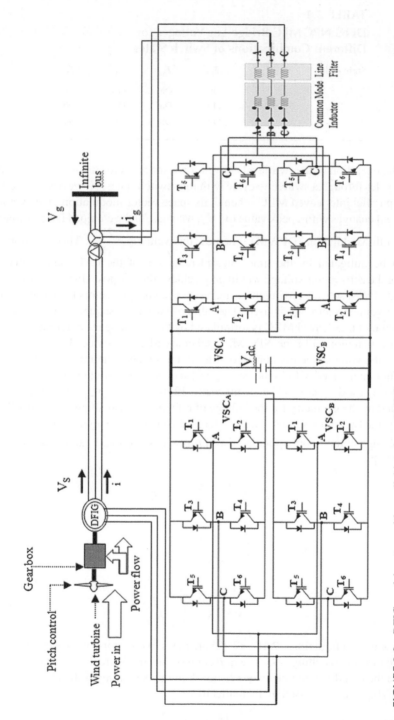

FIGURE 2.8 DFIG model system with parallel interleaved MLC scheme.

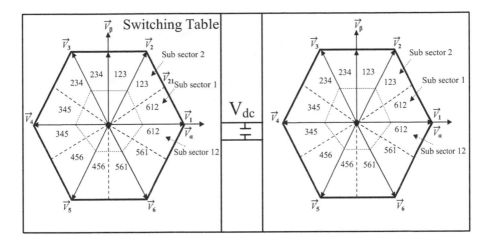

FIGURE 2.9 Switching table for the parallel interleaved MLC scheme.

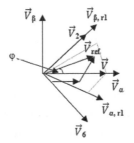

FIGURE 2.10 Space vector reference formation for parallel interleave MLC scheme.

$$V_{\beta,rl} = \frac{1}{2}V_\alpha - \frac{\sqrt{3}}{2}V_\beta \qquad (2.14)$$

$$T_1 = \frac{2}{\sqrt{3}}\frac{\left|\vec{V}_{ref}\right|}{V_{dc}}T_s \sin\left(60° - \varphi\right) \qquad (2.15)$$

$$T_2 = \frac{2}{\sqrt{3}}\frac{\left|\vec{V}_{ref}\right|}{V_{dc}}T_s \sin\varphi \qquad (2.16)$$

$$T_3 = T_6 = \left(T_S - T_1 - T_2\right)/2 \qquad (2.17)$$

The active vectors $\left(T_1, T_2\right)$ adjacent dwell time for the PWM zero state and for the traditional space vector modulation in Equations (2.15)–(2.17) are the same. Though in this case, the $(\vec{V}_3, \vec{V}_6)$ voltage vectors, which are active opposing two near vectors, are used instead of the zero vectors. Consequently, operation range of modulation $(0 \le M \le 2/\sqrt{3})$ is obtained.

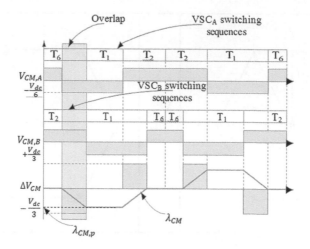

FIGURE 2.11   Switching sequence for parallel interleave MLC scheme.

The DFIG parallel interleaved MLC switching sequence and CM voltages are shown in Figure 2.11 for both converter sides. The converter's two halves are tagged A and B, respectively. With a voltage common to each voltage source converter, the maximum individual VSCs CM voltage is constrained to $\pm\frac{V_{dc}}{6}$.VCM, A and VCM, B and their difference ΔVCM is shown in Figure 2.11. Since only active vectors are used in this topology, the peak CM voltage of the individual VSCs is constrained to $\pm\frac{V_{dc}}{6}$. The CM voltages polarity are equal but opposite for the power converters scheme. Due to the individual CM voltages opposite polarity, the combined use of $\vec{V}_1$ and $\vec{V}_2$ in VSC$_A$ and VSC$_B$, respectively, and the other configuration in sub-sector 1, makes the change in the common mode ΔVCM to confine its value to $\pm\frac{V_{dc}}{3}$. Consequently, the flux linkage for a switching $\lambda_{CM,p}$ depend on the voltage vectors' time of overlap, as shown in Figure 2.11, for the voltage of a particular DC-link. These voltage vectors $\vec{V}_1$ and $\vec{V}_6$ overlap time in $\vec{V}_{ref}$ sub-sector 12 shown in the table of switching, controls the $\lambda_{CM,p}$ value, which varies with the voltage vector $\vec{V}_{ref}$ reference. Because the changes in the reference voltage are with respect to the space vector angle φ, different values of the switching cycle for the power converter are obtained.

2.6   SIMULATION RESULTS AND DISCUSSIONS

2.6.1   SDBR Switching and Position

Simulations were done based on the optimal switching signal (Figure 2.2) and placement (Figure 2.1) of SDBR having a value of 0.01 pu for a severe 3LG fault which occurs at 0.1 s. The 2-level inverter scheme having a DC chopper was used in this investigation. As seen from Figures 2.12–2.14 the following wind generator variables, DC-link voltage, active power and reactive power show improved performance

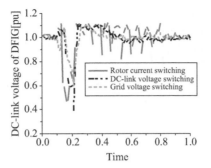

FIGURE 2.12   DC-link voltage of DFIG.

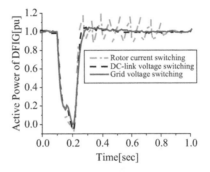

FIGURE 2.13   Active power of DFIG.

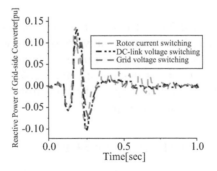

FIGURE 2.14   Grid side reactive power of DFIG.

with low distortions, with the use of the braking resistor, when subjected to transient conditions. This was mainly due to the fact that most of the harmonics and vibrations experienced during the transient period were transferred by the rotor current and DC-link voltage signals to the braking resistor circuitry, compared to when the grid voltage signal was employed. The scenario where the braking resistor was inserted at the stator side of the wind generator, considering the grid voltage switching signal, led to enhanced performance of the terminal voltage and rotor speed variables of the wind

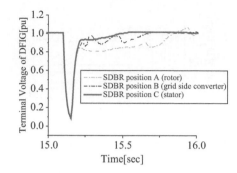

FIGURE 2.15   Terminal voltage of DFIG.

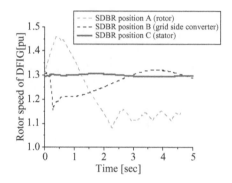

FIGURE 2.16   Rotor speed of DFIG.

generator (Figures 2.15 and 2.16). The reason for this is that the wind generator's active power was enhanced and, consequently, its stability was improved. Besides, the braking resistor limits the rotor speed at times of transient, thus, improving the power output of the wind turbine. Consequently, the recovery of the variables of the wind generator after fault is better because of the control capability of the braking resistor in the acceleration of the rotor speed.

2.6.2   2-LEVEL INTERLEAVED INVERTER AND SDBR

A comparison was made between the single 2-Level Converter (2LC) system and when it is interleaved for the wind turbine system. As seen from Figures 2.17 and 2.18 for the DC-link voltage and reactive power of the wind generator, interleaving the two side converters using the 2LCs enhances the stability of the wind turbine during transient by providing more reactive power within the DFIG power converters (Figure 2.18). Based on the best SDBR performance for the stator side, as given in Section 2.6.1 of the simulation results, a further investigation was carried out for the 2-level interleave inverter system to improve the stability of the wind turbine for the DC-link voltage and the rotor speed variables shown in Figures 2.19 and 2.20,

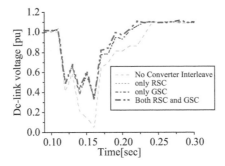

FIGURE 2.17   DC-link voltage of DFIG.

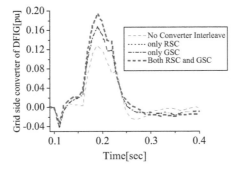

FIGURE 2.18   Grid side converter of DFIG.

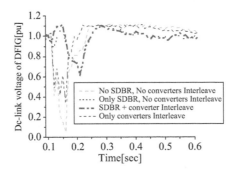

FIGURE 2.19   DC-link voltage of DFIG.

respectively. The DC-link charging current of the capacitor is mitigated during fault scenario, in order to avoid the undervoltage or overvoltage experienced, as shown in Figure 2.19. In addition, this topology effectively controls the voltage and current in the rotor circuitry. This would definitely improve the rotor speed of the DFIG, as shown in Figure 2.20.

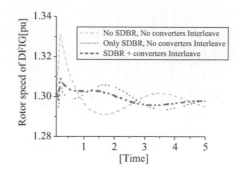

FIGURE 2.20   Rotor speed of DFIG.

2.6.3   3-LEVEL INVERTER, 2-LEVEL INTERLEAVED INVERTER AND SDBR

Investigation was further carried out by comparing the responses of the 3-level inverter and the 2-level interleaved inverter. The responses of the DFIG DC-link voltage and active power in Figures 2.21 and 2.22, respectively, are better for the 2-level interleaved inverter than the 3-level inverter due to more reactive power supplied, as shown in Figure 2.23. However, the response of the rotor current is almost the same

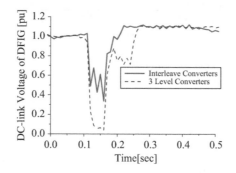

FIGURE 2.21   DC-link voltage of DFIG.

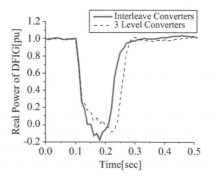

FIGURE 2.22   Real power of DFIG.

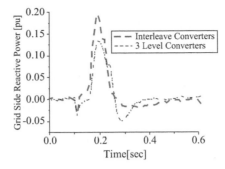

FIGURE 2.23 Grid side converter of DFIG.

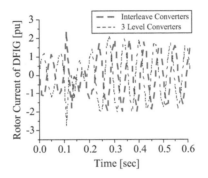

FIGURE 2.24 Rotor current of DFIG.

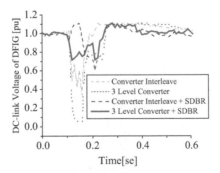

FIGURE 2.25 DC-link voltage of DFIG.

but in a different direction, as shown in Figure 2.24, for both converter systems. An addition of SDBR to both systems, as shown in Figures 2.25–2.28, indicates that the SDBR can further enhance the performance of both inverter schemes, for the DC-link voltage, real power, reactive power of the grid side converter and rotor speed of the wind generator. The response of the 2-level interleave inverter is slightly better because the SDBR limits the reactive power even more (Figure 2.27) for the 3-level inverter.

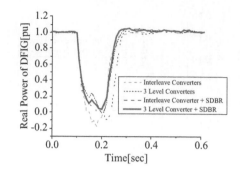

FIGURE 2.26   Real power of DFIG.

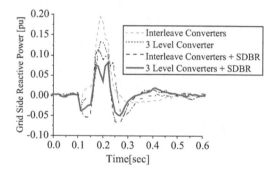

FIGURE 2.27   Grid side converter of DFIG.

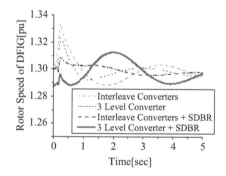

FIGURE 2.28   Rotor speed of DFIG.

2.6.4 PROPOSED PLL CONTROL STRATEGY AND THE VARIOUS SCHEMES

The conventional PLL was used in the earlier analysis of the DFIG wind turbine in this chapter. This section investigates five different schemes for the DFIG wind turbine. Scheme 1 is the 2LC using the conventional PLL, scheme 2 is the 2LC with the proposed PLL scheme in Figure 2.6 and scheme 3 is the 2LC with the proposed PLL

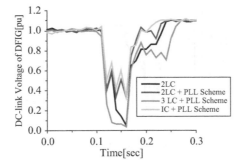

FIGURE 2.29   DC-link voltage of DFIG.

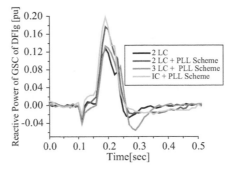

FIGURE 2.30   Reactive power of DFIG.

scheme and SDBR. Schemes 4 and 5 use the 3-Level Converter (3LC) and parallel Interleave Converter (IC), having the proposed PLL scheme and SDBR, respectively. Figure 2.29 shows the response of the DFIG wind turbine DC-link voltage considering the various converter schemes without SDBR. The response of the DC-link voltage using the proposed PLL scheme for the 2LC system is better than when the conventional PLL scheme was used. This is because the proposed PLL scheme helps in enhancing the reactive power of the DFIG grid side converter (Figure 2.30). The coordinated control of SDBR with the proposed scheme shows that better performance of the wind turbine variables could be achieved during transient for the DC-link, real power and rotor speed of the wind generator, as shown in Figures 2.31–2.33. In Figure 2.31, the SDBR and the proposed PLL scheme enhance the single-step 2-level inverter, thus, giving a better response than the interleaved and 3-level inverter schemes. This is also the same for the wind turbine real power in Figure 2.32 and rotor speed in Figure 2.33, respectively.

2.7   CHAPTER CONCLUSION

In this chapter, the DFIG variable speed wind turbine performance operating with three types of inverter system was investigated. The 2-level single-step inverters, 2-level parallel interleaved inverters and 3-level inverter systems for a DFIG wind

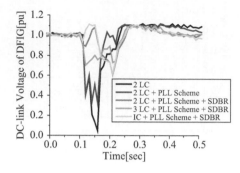

FIGURE 2.31 DC-link voltage of DFIG.

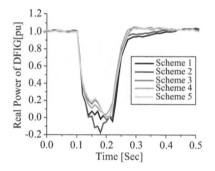

FIGURE 2.32 Real power of DFIG.

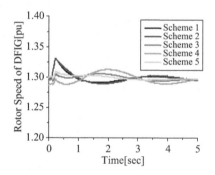

FIGURE 2.33 Rotor speed of DFIG.

turbine during transient were presented. A SDBR was used to enhance the performance of all the inverter system.

Furthermore, a PLL was proposed in conjunction with the SDBR system to improve the performance of the wind turbine using a 2-level single-step inverter system. It was observed that the proposed PLL SDBR strategy can further improve the performance of the 2-level single-step inverter system during transient, thus enhancing the transient responses of the wind turbine.

REFERENCES

[1] L. Xu, and P. Cartwright, "Direct active and reactive power control of DFIG for wind energy generation," *IEEE Transactions Energy Conversion*, vol. 21, no. 3, pp. 750–758, 2006.

[2] K. E. Okedu, "Improving the transient performance of DFIG wind turbine using pitch angle controller low pass filter timing and network side connected damper circuitry," *IET Renewable Power Generation*, vol. 14, no. 7, pp. 1219–1227, 2020.

[3] G. Gohil, L. Bede, R. Teodorescu, T. Kerekes, and F. Blaabjerg, "An integrated inductor for parallel interleaved VSCs and PWM schemes for flux minimization," *IEEE Transactions on Industrial Electronics*, vol. 62, no. 12, pp. 7534–7546, 2015.

[4] H. Wang, M. Liserre, and F. Blaabjergy, "Toward reliable power electronics-challenges, design tools and opportunities," *IEEE Industrial Electronics Magazine*, vol 7, no. 2, pp. 17–26, 2013.

[5] K. E. Okedu, "Hybrid control strategy for variable speed wind turbine power converters," *International Journal of Renewable Energy Research*, vol. 3, no. 2, pp. 283–288, 2013.

[6] A. Volke, and M. Hornkamp, *IGBT Modules-Technologies, Driver and Applications*, 2nd ed., München, Germany: Infineon Technologies, AG, 2012. ISBN: 978-3-00-040134-3.

[7] T. Ohi, A. Iwata, and K. Arai, "Investigation of gate voltage oscillations in IGBT module under short circuit conditions," *Proceedings of 33rd Annual Power Electronics Specialist Conference*, vol. 4, pp. 1758–1763, 2002.

[8] K. E. Okedu, S. M. Muyeen, R. Takahashi, and J. Tamura, "Wind farms fault ride through using DFIG with new protection scheme," *IEEE Transactions on Sustainable Energy*, vol. 3 no. 2, pp. 242–254, 2012.

[9] K. E. Okedu, S. M. Muyeen, R. Takahashi, and J. Tamura, "Improvement of fault ride through capability of wind farm using DFIG considering SDBR," *14th European Conference of Power Electronics EPE*, Birmingham, UK, pp. 1–10, 2011.

[10] K. E. Okedu, "Enhancing the performance of DFIG variable speed wind turbine using parallel integrated capacitor and modified modulated braking resistor," *IET Generation Transmission & Distribution*, vol. 13, no. 15, pp. 3378–3387, 2019.

[11] G. Rashid, and M. H. Ali, "Nonlinear control-based modified BFCL for LVRT capacity enhancement of DFIG based wind farm," *IEEE Transactions on Energy Conversion*. doi: 10.1109/TEC.2016.2603967, No. 99, 2016.

[12] M. Firouzi, and G. B. Gharehpetian, "LVRT performance enhancement of DFIG-based wind farms by capacitive bridge-type fault current limiter," *IEEE Transactions on Sustainable Energy*, vol. 9, No. 3, pp. 1118–1125, 2018.

[13] L. Chen, C. Deng, F. Zheng, S. Li, Y. Liu, and Y. Liao, Fault ride-through capability enhancement of DFIG-based wind turbine with a flux-coupling-type SFCL employed at different locations. *IEEE Transactions on Applied Superconductivity*, vol. 25, Article 5201505, 2015.

[14] PSCAD/EMTDC Manual, Version 4.6.0; Manitoba HVDC Lab.: Winnipeg, MB, Canada, 2016.

[15] K. E. Okedu, and H. Barghash, "Investigating variable speed wind turbine transient performance considering different inverter schemes and SDBR," Frontiers in Energy Research-Smart Grids, vol. 8, pp. 1–16, Article 604338, 2021, doi: 10.3389/fenrg.2020.604338.

[16] L. You-Wei, C. Zhi-Hui, and S. Juan, "Application of PLL in the generator side converter for doubly fed wind power generation systems," *Energy Procedia*, vol. 16, pp. 1822–1830, 2012.

[17] J. L. Da Silva, R. G. de Oliveira, S.R. Silva, B. Rabelo, and W. Hofmann, "A discussion about a start-up procedure of a doubly-fed induction generator system," *NORPIE/2008, Nordic Workshop on Power and Industrial Electronics*, 9–11 June 2008.

[18] P. Jung-Woo, L. Ki-Wook, K. Dong-Wook, L. Kwang-Soo, and P. Jin-Soon, "Control method of a doubly-fed induction generator with a grid synchronization against parameter variation and encoder position," *Industry Applications Conference, 2007, 42nd IAS Annual Meeting, Conference Record of 2007 IEEE*, New Orleans, LA, 23–27 September 2007, pp. 931–935.

[19] R. Takahashi, J. Tamura, M. Futami, M. Kimura, and K. Idle, "A new control method for wind energy conversion system using doubly fed synchronous generators," *IEEJ Transactions on Power and Energy*, vol. 126, no. 2, pp. 225–235, 2006.

[20] K. E. Okedu, "Enhancing DFIG wind turbine during three-phase fault using parallel interleaved converters and dynamic resistor," *IET Renewable Power Generation*, vol. 10, no. 6, pp. 1211–1219, 2016.

3 DFIG Performance and Excitation Parameters

3.1 CHAPTER INTRODUCTION

This chapter tends to improve the performance of the DFIG-based wind turbine, by investigating the effects of the excitation parameters of the power converter Insulated Gate Bipolar Transistors (IGBTs) of the DFIG wind turbine during transient conditions. The excitation parameters of the IGBTs investigated are the turn-on and turn-off resistances, forward break over and reverse withstand voltages. An extensive analysis of the effects of the excitation parameters of the IGBTs was carried out considering 15 scenarios. The IGBT turn-on resistance has a considerable effect on the responses of the wind generator's variables, while its turn-off resistance has little or minimal effects on the variables of the wind generator [1]. However, the forward break over and reverse withstand voltages of the IGBT have no effects on the performance of the wind generator variables during transient conditions. The optimal excitation parameters that gave the best responses of the wind generator variables during transient were used to further analyze the performance of the wind generator, by proposing a hybrid coordinated control of a new Phase Lock Loop (PLL) configuration and the SDBR. The proposed PLL control is based on a delayed signal that helps the phase of the positive sequence component to be obtained quickly and accurately, despite the grid voltage being disturbed. Hence, it has a better tracking of the positive components of the grid voltage, compared to the conventional PLL. The SDBR mitigates the overvoltage experienced in the rotor circuitry and the rotor high currents. Thus, the wind generator's power converter is intact during operation because these actions would eliminate the capacitor's DC-link high charging current, which is dangerous to the power converters. In this chapter, optimum size and placement of the braking resistor in the wind generator were used in enhancing the wind generator during transient, in conjunction with the proposed PLL scheme for the DFIG wind generator, as earlier discussed in Chapter 2 of this book. In addition, a comparative study using the proposed hybrid scheme and the conventional PLL DFIG scheme was presented. Results obtained using the proposed control strategies for the DFIG during severe three-phase fault scenario were better than those using the conventional PLL scheme. Simulations were run in PSCAD/EMTDC [2].

3.2 MODEL SYSTEM OF STUDY

The model system of study is shown in Figure 3.1. The DFIG wind turbine modeling is already presented in Chapter 1. The best SDBR position, switching signal and control strategy were evaluated considering the most severe fault scenario of 3 LG in Figure 3.1, model system [3]. The optimum size of the SDBR, duration of operation

DOI: 10.1201/9781003350910-3

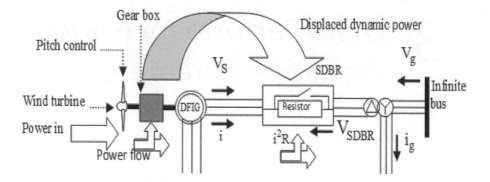

FIGURE 3.1   DFIG model with SDBR dynamics.

TABLE 3.1
Parameters of DFIG Wind Turbine

Parameters ratings

Rated power	2 MW
Rated voltage	690 V
Stator resistance	0.01 pu
Stator leakage reactance	0.15 pu
Magnetizing reactance	3.5 pu
Rotor resistance	0.01 pu
Rotor leakage reactance	0.15 pu
Inertia constant	1.5 s

and best switching signals were investigated already in [4, 5]. The best SDBR size of 0.05 pu, with grid voltage switching signal topology for the DFIG during transient conditions, was used for further investigation, considering the proposed PLL scheme. The parameters of the DFIG wind generator are given in Table 3.1.

3.3   DFIG VARIABLE SPEED DRIVE CONTROL

Figure 3.2 shows the classical configuration and power flow of a wind generator based on DFIG technology, with bidirectional power converters. The utilization of the voltage source converters in an economical way is one of the important features of this type of wind turbine. This is because unlike the synchronous wind generators that employ full converter rating, only the power electronic rotor circuitry is required in the case of DFIG wind turbines. During normal operation, the optimal power of the wind generator is achieved considering the conditions of the wind power. The maximum power point tracking characteristics of the wind generator is directly related to its d-axis reference current. The regulation of active power of the wind generator is possible via the control of the pitch angle. This is done by frequency control and grid support based on the provision of power margins. From Figure 3.2, the DC-link voltage is maintained constant at 1.0 pu, by the regulation of the quadrature axis current,

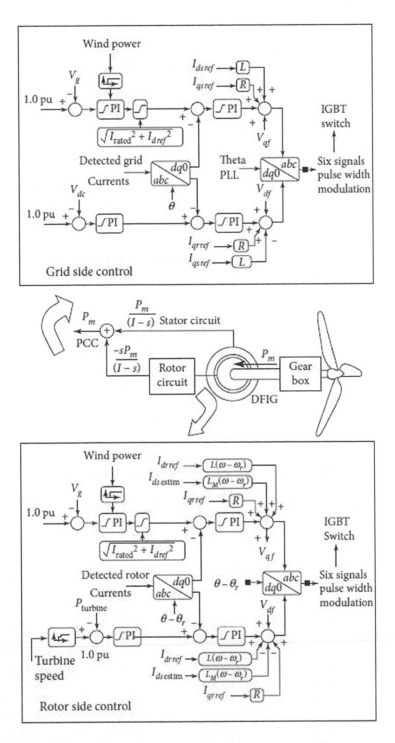

FIGURE 3.2 DFIG control strategy.

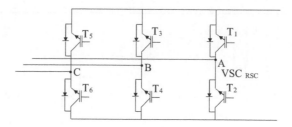

FIGURE 3.3   DFIG power converter with IGBTs.

while the grid voltage is maintained constant at 1.0 pu, by the regulation of the direct axis current. Reactive power dissipation is achieved by the power converters of the wind generator by regulating the grid voltage.

3.4   INSULATED GATE BIPOLAR TRANSISTOR EXCITATION PARAMETERS

The schematic diagram of the rotor side converter of the DFIG variable speed wind turbine is shown in Figure 3.3, with the IGBTs connected. Table 3.2 gives the summary of the excitation parameters and their respective ratings, investigated in this chapter.

The turn-on and turn-off resistances, forward break over and reverse withstand voltages excitation parameters of the IGBTs were investigated considering the various scenarios shown in Table 3.2. The IGBT turn-on resistance has a considerable effect on the responses of the wind generator's variables, while its turn-off resistance has little or minimal effects on the variables of the wind generator. These would be

TABLE 3.2

Excitation Parameters of the Insulated Gate Bipolar Transistors

Scenario	IGBT turn-on resistance (Ohm)	IGBT turn-off resistance (Ohm)	Forward break overvoltage (kV)	Reverse withstand voltage (kV)
1	0.001	1.0×10^6	1.0×10^5	1.0×10^5
2	0.002	1.0×10^6	1.0×10^5	1.0×10^5
3	0.003	1.0×10^6	1.0×10^5	1.0×10^5
4	0.001	1.0×10^7	1.0×10^5	1.0×10^5
5	0.001	1.0×10^8	1.0×10^5	1.0×10^5
6	0.002	1.0×10^7	1.0×10^5	1.0×10^5
7	0.002	1.0×10^8	1.0×10^5	1.0×10^5
8	0.003	1.0×10^7	1.0×10^5	1.0×10^5
9	0.003	1.0×10^8	1.0×10^5	1.0×10^5
10	0.001	1.0×10^6	1.0×10^6	1.0×10^6
11	0.001	1.0×10^6	1.0×10^7	1.0×10^7
12	0.002	1.0×10^6	1.0×10^6	1.0×10^6
13	0.002	1.0×10^6	1.0×10^7	1.0×10^7
14	0.003	1.0×10^6	1.0×10^6	1.0×10^6
15	0.003	1.0×10^6	1.0×10^7	1.0×10^7

demonstrated in the simulation results and discussion section of this chapter. The forward break over and reverse withstand voltages of the IGBT have no effect on the performance of the wind generator variables, during transient conditions. Thus, the variation of the forward break over and reverse withstand voltages of the wind generator power converter does not contribute to the wind turbine enhancement during steady state and grid fault conditions. The optimal excitation parameters that gave the best responses of the wind generator variables during transient were used for further analysis of the wind turbine considering the new PLL scheme that was proposed in Chapter 2 of this book.

3.5   FAULT RIDE THROUGH REQUIREMENTS FOR WIND FARMS

Wind farms are faced with drops in voltage during fault scenarios. Since the grid voltage depends on reactive power injection, recent grid codes require immediate restoration of operation of the wind farms after grid faults, as shown in Figure 3.4 [6, 7]. The grid codes stipulate that within the time frame shown in the figure, the collapse in voltage of the wind farm must be regulated to normal, for the wind farm to stay in the grid. Otherwise, the wind farm must be shut down or disconnected.

3.6   EVALUATION OF THE SYSTEM PERFORMANCE

3.6.1   EFFECTS OF THE INSULATED GATE BIPOLAR EXCITATION PARAMETERS ON THE DFIG WIND TURBINE

The effects of the excitation parameters on the DFIG variable speed wind turbine responses considering a severe bolted three-phase-to-ground fault were investigated

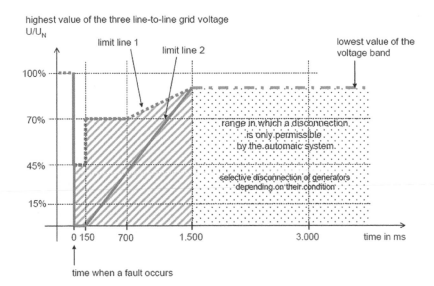

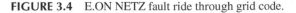

FIGURE 3.4   E.ON NETZ fault ride through grid code.

in this section based on the different scenarios given in Table 3.2. Some of the simulation results of the DFIG variables are presented in Figures 3.5–3.17.

Figure 3.5 shows the response of the DFIG wind turbine DC-link voltage for the first three scenarios of Table 3.2. With a too-high IGBT turn-on resistance of 0.003 Ohms, the DC-link variable of the wind generator did not recover on time, while a too-low resistance of 0.001 Ohms gives a faster recovery of the wind generator DC-link voltage. However, there is an overshoot of the DC-link variable considering the

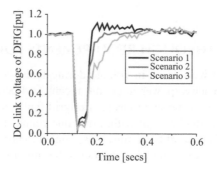

FIGURE 3.5   DC-link voltage of DFIG.

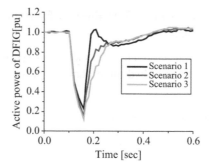

FIGURE 3.6   Active power of DFIG.

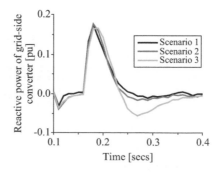

FIGURE 3.7   Reactive power of DFIG grid side converter.

low resistance of 0.001 Ohms. Thus, a moderate turn-on resistance value of size 0.002 Ohms for scenario 2 gives a better performance for the DC-link voltage, the active power (Figure 3.6) and the reactive power of the wind generator grid side converter shown in Figure 3.7. Though scenario 1 tends to have a better performance of the rotor speed response of the wind generator during transient, as shown in Figure 3.8; however, the response of the terminal voltage of the wind generator, shown in Figure 3.9, for this scenario is lower than that of scenario 2. Thus, a moderate and

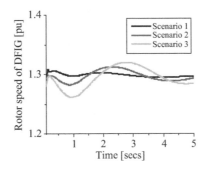

FIGURE 3.8 Rotor speed of DFIG.

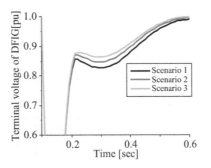

FIGURE 3.9 Terminal voltage of DFIG.

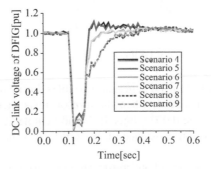

FIGURE 3.10 DC-link voltage of DFIG.

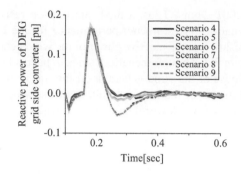

FIGURE 3.11   Reactive power of DFIG grid side converter.

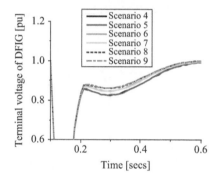

FIGURE 3.12   Terminal voltage of DFIG.

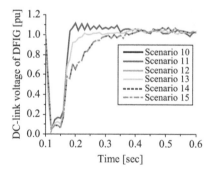

FIGURE 3.13   DC-link voltage of DFIG.

better response of the DFIG wind generator variables during a severe three-phase-to-ground fault could be achieved, when the turn-on resistance of the IGBTs of the power converter is of value 0.002 Ohms. It should be noted that, in the investigation of the effects of the turn-on resistance of the IGBT, the other excitation parameters were kept constant.

A further analysis was carried out considering scenarios 4–9 in Table 3.1, by varying the turn-on and turn-off resistances of the IGBTs, while keeping constant the

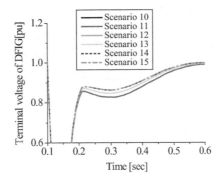

FIGURE 3.14   Terminal voltage of DFIG.

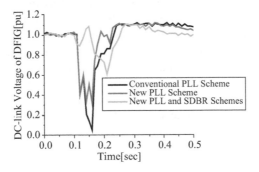

FIGURE 3.15   DC-link voltage of DFIG.

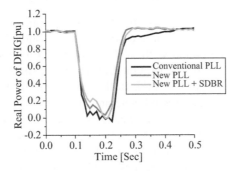

FIGURE 3.16   Real power of DFIG.

forward break over voltage and reverse withstand voltage as shown in Figures 3.10–3.12, respectively. From Figures 3.10 to 3.12, the turn-off resistance of the IGBTs of the wind generator power converter has little or no effects on the responses of the wind generator's DC-link, reactive power and terminal voltage during transient conditions.

To further investigate the effects of the excitation parameters of the IGBTs of the wind generator power converters for the DFIG wind turbine, scenarios 10–15 in

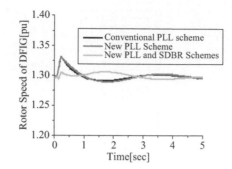

FIGURE 3.17   Rotor speed of DFIG.

Table 3.1 were considered by varying the forward break over voltage and reverse withstand voltage, while keeping the turn-off resistance of the IGBTs constant, as shown in Figures 3.13 and 3.14, respectively.

From Figures 3.13 and 3.14, the forward break over voltage and reverse withstand voltage of the power converter of the IGBTs do not have any effects on the DC-link voltage and terminal voltage variables of the DFIG wind turbine. In light of the above, only the IGBT's turn-on resistance has a great impact on the variables of the wind generator. Thus, the optimal value of 0.002 Ohms turn-on resistance and the IGBTs excitation parameters considering scenario 2 of Table 3.2, would be used for further analysis of the new PLL scheme proposed for the DFIG wind turbine in the subsequent section of this chapter.

3.7   PROPOSED PLL CONTROL STRATEGY AND SDBR SCHEME

Based on the optimal performance of the excitation parameters for the power converter IGBTs of the DFIG wind turbine presented in Section 3.6.1, a further investigation was carried out to improve the stability of the DFIG wind turbine using the new PLL scheme, in conjunction with the SDBR control strategy. Simulations were run for three scenarios in this section. In the first scenario, the conventional PLL scheme was used for the DFIG wind generator, while the second scenario uses the new PLL scheme for the DFIG wind generator. The combination of the proposed PLL scheme and best control strategies of the SDBR was used in the third scenario to improve the stability of the wind generator during transient condition. Some of the DFIG variables for these analyses are shown in Figures 3.15–3.17, for a three-phase-to-ground severe fault.

Figure 3.15 shows the response of the DFIG wind turbine DC-link voltage for the conventional PLL scheme and the new PLL scheme. It could be observed from the figure that the proposed PLL scheme helps in enhancing the performance of the DC-link voltage during transient with the help of the delay element, as discussed earlier. A further investigation of using the new PLL scheme and the best working conditions of the SDBR improved the variables of the wind generator during transient condition. The reasons for this improved performance are based on the benefits of the SDBR earlier given for the limitation of the rotor overvoltage and current of the rotor

TABLE 3.3
Comparison of the Proposed PLL Scheme and Other PLL Solutions for DFIG

	Proposed PLL scheme	Traditional PLL scheme
1	Has delay elements in the feedback system to the PI controller. Delayed feedback elements help in enhancing the performance of the wind generator during transient.	Does not have this feature.
2	It counteracts the disturbances of twice the frequency of the grid by combining the dq components by ¼ period interleaving delays, with respect to the original period.	The positive sequence converts to DC components while negative sequence component to a value that is twice the frequency component.
3	It helps in mitigating the influence of the negative sequence components.	Can only be used to remove the steady state error, thus the negative voltage will have an influence on its output.
4	It considers the rated line-to-line voltage, in addition to the induced stator voltages for effective synchronization.	Considers the induced stator voltages to generate offset angle compensator, for phase shift.
5	Effectively detect and track faster the positive and negative sequence components of a disturbed grid voltage. Helps the phase of the positive sequence component to be obtained quickly and accurately, despite the grid voltage being disturbed.	Tracking of the positive and negative sequence variables and components during transient is quite slow.

circuitry of the wind generator, as shown in Figure 3.15. As a result, the real power and rotor speed are enhanced in Figures 3.16 and 3.17. The wind turbine power can be improved using the braking resistor because the rotor speed can be limited with few oscillations during fault scenarios, as shown in Figure 3.17. Therefore, the combination of the proposed PLL scheme and the best working conditions and control strategy of the SDBR will improve the post-fault recovery of the DFIG wind turbine system variables. The comparison of the proposed PLL scheme to traditional PLL solutions in the literature is given in Table 3.3.

3.8 CHAPTER CONCLUSION

The performance of a DFIG variable speed wind turbine using a coordinated control strategy of the optimal excitation parameters of the Insulated Gate Bipolar Transistors (IGBTs), a new PLL and Series Dynamic Braking Resistor (SDBR), during grid fault has been investigated in this chapter. The responses of the DFIG variables during transient conditions using the hybrid control strategies of the excitation parameters of the IGBTs, SDBR and the new PLL scheme were better than the conventional PLL scheme. The IGBTs turn-on resistance has a lot of impact on the DFIG variables during transient compared to the other excitation parameters such as the turn-off resistance, forward break over voltage and reverse withstand voltage. When the turn-on resistance is too high or low, the variables of the wind generator cause overshoot or are too slow to recover after the grid fault. A well sized and positioned SDBR in the stator side of the wind generator was used to enhance the performance of the new

DFIG PLL scheme. As part of future scope, an optimization approach using Dragon Fly Algorithm is the next phase of this work for optimal performance of the wind generator.

REFERENCES

[1] K. E. Okedu, and H. Barghash, "Enhancing the performance of DFIG wind turbines considering excitation parameters of the insulated gate bipolar transistors and a new PLL scheme," *Frontiers in Energy Research-Smart Grids*, vol. 8, pp. 1–14, Article 620277, 2021, doi: 10.3389/fenrg.2020.620277.

[2] PSCAD/EMTDC Manual, Version 4.6.0; Manitoba HVDC Lab.: Winnipeg, MB, Canada, 2016.

[3] K. E. Okedu, "Optimal position and best switching signal of SDBR in DFIG wind turbine low voltage ride through conditions," *IEEE International Electric Machines and Drives Conference (IEMDC 2017)*, Paper ID 115, Miami, FL, 21–24 May 2017.

[4] K. E. Okedu, S. M. Muyeen, R. Takahashi, and J. Tamura, "Wind farms fault ride through using DFIG with new protection scheme," *IEEE Transactions on Sustainable Energy*, vol. 3, no. 2, pp. 242–254, 2012.

[5] K. E. Okedu, "Determination of the most effective switching signal and position of braking resistor in DFIG wind turbine under transient conditions," *Electrical Engineering, Springer*, vol. 102, no. 11, pp. 471–480, 2020.

[6] E.ON NETZ GmbH, Grid Connection Regulation for High and Extra High Voltage. E. ON NETZ GmbH, Essen, Germany, 2006.

[7] M. K. Ghartemani, S. A. Khajehoddin, P. K. Jain, and A. Bakhshai, "Problems of startup and phase jumps in PLL systems," *IEEE Transactions on Power Electronics*, vol. 27, no. 4, pp. 1830–1838, 2012.

4 PMSG Performance and Excitation Parameters

4.1 CHAPTER INTRODUCTION

Based on the recent emerging grid requirements, wind farms are expected to have good performance under grid fault, considering the capability of voltage control. There are bound to be intricacies in operating the power grid smoothly, due to grid disturbances. Consequently, it is necessary to proffer solutions using new techniques to overcome this shortcoming, considering the complex nature of the control strategies of modern power grids [1]. The topologies of Variable Speed Wind Turbine (VSWT) have various generator and converter schemes with regard to capturing of energy, cost, efficiency and control strategies complexity [2]. The two most commonly used VSWTs for wind energy conversion in modern wind farms are the Doubly Fed Induction Generator (DFIG) wind turbine and the Permanent Magnet Synchronous Generator (PMSG)-based VSWT. Though the earlier wind turbine is more popular, the later wind turbine technology has a more feasible wind generation technology, making it more promising due to its self-excited nature; hence, it is possible to operate higher efficiency and power factor. Besides, the PMSG technology has no gearbox system because of its low rotational speed. Therefore, no careful and regular maintenance is required in this wind turbine topology, unlike the DFIG-based wind turbine [3]. More so, the PMSG wind turbine power converters have room for flexible control of active and reactive power dissipation during transient conditions [4, 5]. This type of wind turbine technology has a full-rated back-to-back power converter tied to the grid. As a result, compared to the earlier wind turbine topology, the flexibility in the later type of wind turbine is maximum, making the control of real and reactive power more effective. However, some challenges of the PMSG wind turbine are high cost, construction and control strategy complexities.

In the literature, there are many reports on the control strategies and fault ride through topologies concerning PMSG wind turbines. Nasiri and Mohammadi [6] enhanced the PMSG wind turbine considering its peak current limitation, while a control strategy based on the Maximum Power Point Tracking (MPPT) for both power converters was used in [7]. Superconducting Fault Current Limiter (SFCL) was used in [8] to improve the performance of the PMSG wind turbine during grid fault. The use of expensive Static Synchronous Compensator (STATCOM) in [9] and DC braking chopper Fault Ride Through (FRT) solution in [10–12], are some of the most common control strategies employed in enhancing the performance of the PMSG wind turbine during transient state.

Furthermore, soft computing techniques have been extended to the PMSG wind turbine control topology. In [13], the fuzzy logic controller was employed for

DOI: 10.1201/9781003350910-4

the PMSG, with some benefits over conventional controllers of the wind generator. The drawback of this computing scheme is the architecture and low sensitivity level to parameter variation. A combined approach of using conventional Proportional Integral Derivative (PID) controllers with the soft computing scheme of fuzzy logic was carried out in [4], for improved performance of the PMSG wind generator. Some more complex soft computing techniques were extended for the PMSG wind turbine by implementing linear observer extended state in [14] and fuzzy adaptive strategy in [15], respectively.

A dynamic voltage restorer was used to improve the performance of the PMSG wind generator in [16]. In the literature, an SFCL that is modified considering flux-coupling was used to enhance the FRT performance of the PMSG wind turbine. The principle and theoretical influence of the modified SFCL structure on the PMSG ride through capability were conducted and a comparison of the SFCL with a dynamic braking chopper was performed in [17]. In [18], a resistive SFCL was employed as an additional self-healing mechanism to support large wind power plant, in order to enhance the rated active power of the PMSG wind turbine. The scheme was also able to improve the DC-link voltage smoothness and the Low Voltage Ride Through (LVRT) capability of the wind generator. A cooperative strategy that is integrated with a cost-effective Superconducting Magnetic Energy Storage (SMES) unit, considering two modified wind turbine generator control strategies was presented in [19]. In that study, the effective utilization of SMES system was achieved during grid faults based on the overvoltage-suppressing effect in the DC-link of the wind generator. A new Multi-Step Bridge-type Fault Current Limiter (MSBFCL) PMSG FRT enhancement was presented in [20]. The multi-resistors were connected in parallel with the Insulated Gate Bipolar Transistor (IGBT) switches and were employed to provide a controllable discrete step resistance.

Though these enhancement schemes highlighted above were able to improve the performance of the PMSG wind turbine during transient state, however, they are additional circuitries to the PMSG wind turbines. In this regard, this chapter tends to propose a simple and cost-effective FRT solution for the PMSG wind generator. The effects of the excitation parameters of the power converter's IGBTs of the PMSG with turbine during transient condition was investigated. The excitation parameters of the IGBTs that were investigated include the turn-on and turn-off resistances. It has already been reported in [21, 22] that the forward break over and reverse withstand voltages have not much effect on the transient performance of the variable speed wind generators. An extensive analysis of the effects of the excitation parameters of the IGBTs was carried out by considering three scenarios of the IGBT's turn-on resistances with and without considering DC-chopper braking resistor for overvoltage protection of the PMSG wind turbine. The IGBT turn-on resistance has a significant effect on the responses of the wind generator's variables, while its turn-off resistance has little or minimal effects on the variables of the wind generator. The results obtained using the PMSG wind turbines were compared to those obtained using the DFIG wind turbines. The evaluation of the system performance was carried out using Power System Computer Design and Electromagnetic Transient Including DC (PSCAD/EMTDC).

TABLE 4.1
Parameters of the PMSG Wind Turbine

Parameters	Ratings
Rated power	5.0 MW
Stator resistance	0.01 pu
d-axis reactance	1.0 pu
q-axis reactance	0.7 pu
Machine inertia (H)	3.0
Effective DC-link protection	0.2 Ω
Overvoltage Protection System (OVPS)	110%

4.2 MODELING OF THE PMSG WIND TURBINE

The modeling of the PMSG wind turbine can be referred to in Chapter 1 of this book. The parameters of the PMSG wind turbine are given in Table 4.1.

4.3 CONTROL STRATEGY FOR THE PMSG WIND TURBINE

Figure 4.1(a) shows the controller of the Machine Side Converter (MSC) of the PMSG that regulates both active and reactive power of the wind turbine. The abc to dq variables transformation is achieved via angle position rotor (θ_r), considering the rotor speed of the wind generator. The d-axis current (I_{sd}) regulates the active power (P_s), while the q-axis current (I_{sq}) controls the reactive power (Q_s) of the wind generator. The MPPT technique for the characteristic of the wind turbine in Chapter 1 is employed in the reference active power (P_{ref}). Usually, the reference reactive power (Q_S^*) is fixed at 0, to obtain effective operation of unity power factor. Three reference phase voltages (V_{sa}^*, V_{sb}^*, V_{sc}^*) are generated for the PWM switching, through the outputs of the current controller, based on the voltage references V_{sd}^* and V_{sq}^*. The GSC control of the wind generator is also shown in Figure 4.1(b) and is regulated considering the d-q rotating reference frame, based on the grid voltage being the same as the speed of rotation. The Park transformation is used in converting (I_{ga}, I_{gb}, I_{gc}) and (V_{ga}, V_{gb}, V_{gc}) three-phase voltage and current of the grid into their rotating d-q reference frame. The extraction of the phase angle (θ_g) of the grid side is done by considering the phase lock loop.

4.4 EXCITATION PARAMETERS OF THE INSULATED GATE BIPOLAR TRANSISTORS OF THE PMSG WIND TURBINE

The schematic diagram of the basic structure and equivalent circuit of an IGBT switch is shown in Figure 4.2. Table 4.2 gives the summary of the excitation parameters and their respective ratings, investigated in this chapter.

The operation of the IGBT in Figure 4.2 is characterized by the conductivity modulation of the n^- region, in addition to the operation represented by the equivalent

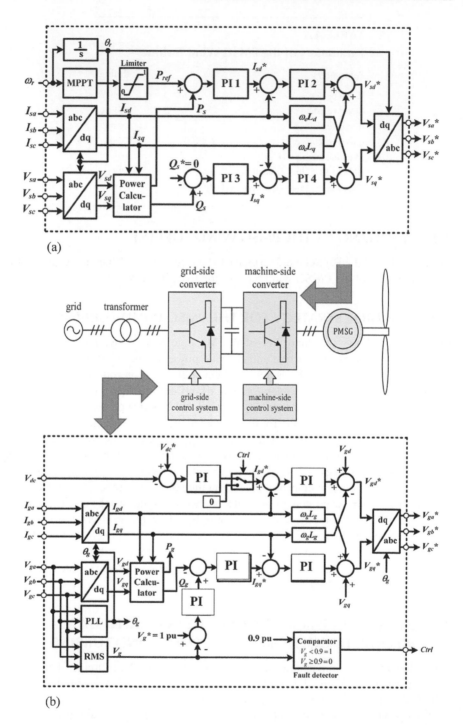

(a)

(b)

FIGURE 4.1 Control strategy of PMSG-based wind turbine. (a) Machine side converter control circuit of the PMSG and (b) grid side converter control circuit of the PMSG.

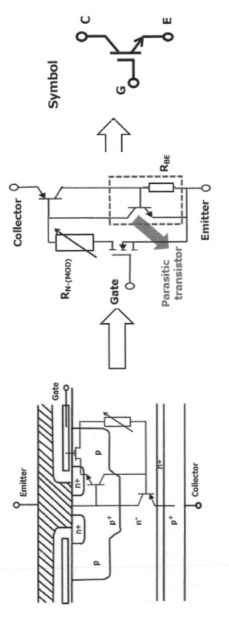

FIGURE 4.2 Basic structure and equivalent circuit of IGBT.

TABLE 4.2

Excitation Parameters of the Insulated Gate Bipolar Transistors

Scenarios	IGBT turn-on resistance (Ohm)	IGBT turn-off resistance (Ohm)	Forward break over voltage (kV)	Reverse withstand voltage (kV)
1	0.001	1.0×10^6	1.0×10^5	1.0×10^5
2	0.002	1.0×10^6	1.0×10^5	1.0×10^5
3	0.003	1.0×10^6	1.0×10^5	1.0×10^5

circuit. In the IGBT circuit, the holes are the minority carriers and because they are injected into the n^- region from the $p^+ - n^+$ region, conductivity modulation occurs in the n^- region. From the equivalent circuit in Figure 4.2, the saturation voltage $\left(V_{CE(sat)}\right)$ of the IGBT can be expressed as [23]:

$$V_{CE(sat)} = V_{BE} + I_D \left(R_{N-(MOD)} + R_{ch} \right) \qquad (4.1)$$

From Equation (4.1):

V_{BE} is the base-emitter voltage

I_D is the drain current

$R_{N-(MOD)}$ is the resistance of the n^- region after conductivity modulation

R_{ch} is the channel resistance.

If the collector current and the DC gain current are I_C and h_{FE}, respectively.

Then, I_D is calculated as:

$$I_D = I_C / h_{FE} \qquad (4.2)$$

The total current of the IGBT $\left(I_{IGBT}\right)$ is:

$$I_{IGBT} = I_D + I_C \qquad (4.3)$$

Equation (4.1) reflects that the saturation voltage $V_{CE(sat)}$ of an IGBT depends on I_D, which is a direct function of h_{FE} in Equation (4.2). From Equation (4.1), increasing the resistance in the n^- region would result in conductivity modulation of the IGBT in the PMSG wind turbine.

The ratings of the turn-on and turn-off resistances, forward break over and reverse withstand voltages excitation parameters of the IGBTs considered in this study for the three scenarios are shown in Table 4.2. The IGBT turn-on resistance has a considerable effect on the responses of the PMSG wind generator's variables, while its turn-off resistance has little or minimal effects on the variables of the wind generator. These would be demonstrated in the simulation results and discussion section of this chapter. The forward break over and reverse withstand voltages of the IGBT have no effect on the performance of the wind generator variables, during transient conditions. Thus, the variation of the forward break over and reverse withstand voltages of the wind generator power converter does not contribute to the wind turbine enhancement during steady state and grid fault conditions.

4.5 EVALUATION OF SYSTEM PERFORMANCE

4.5.1 EFFECTS OF THE INSULATED GATE BIPOLAR EXCITATION PARAMETERS ON THE PMSG WITHOUT OVERVOLTAGE PROTECTION SCHEME (OVPS)

The effects of the excitation parameters on the PMSG variable speed wind turbine responses considering a severe bolted three-phase to ground fault was investigated in this section of this chapter based on the different scenarios given in Table 4.2. Some of the simulation results of the PMSG variables are presented in Figures 4.3–4.7.

The system performance was evaluated using PSCAD/EMTDC [24] environment. The fault type is a severe three-phase of 100 ms happening at 10.1 s, with the circuit breakers operation sequence opening and reclosing at 10.2 s and 11 s, respectively, on the faulted line at the terminals of the PMSG wind turbine. Figure 4.3 shows the response of the PMSG wind turbine DC-link voltage for the three scenarios in Table 4.2. With a high IGBT turn-on resistance of 0.003 Ohms in scenario 3, the DC-link variable of the wind generator recovered on time, compared to a low

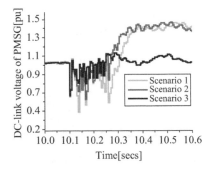

FIGURE 4.3 DC-link voltage of PMSG without OVPS.

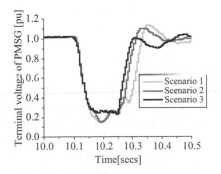

FIGURE 4.4 Terminal voltage of PMSG without OVPS.

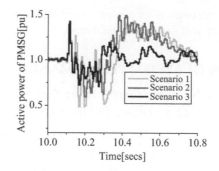

FIGURE 4.5   Active power of PMSG without OVPS.

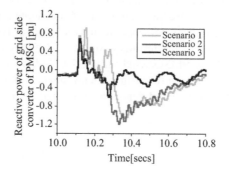

FIGURE 4.6   Reactive power of PMSG without OVPS.

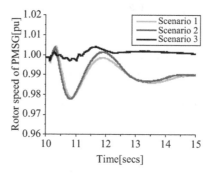

FIGURE 4.7   Rotor speed of PMSG without OVPS.

resistance of 0.001 Ohms for scenario 1 and 0.002 Ohms for scenario 2, respectively. It should be noted that the IGBT's turn-off resistance, forward break over voltage and reverse withstand voltage were kept constant in the course of the study. In Figure 4.4, the terminal voltage for scenarios 1 and 2 gave more voltage overshoot and dip, with less settling time than scenario 3. In Figure 4.5, the response of the active power of the wind generator was better controlled in scenario 3, with less over shot, few oscillations and faster settling time. The PMSG reactive power dissipation was more for scenarios 1 and 2, respectively, in Figure 4.6. While in scenario 3, the higher resistance value of the turn-on resistance was able to limit the reactive power dissipation within the power converter limit. In Figure 4.7, the rotor speed of the wind generator recovered faster for scenario 3, with less overshoot and oscillations, compared to the other scenarios. Scenarios 1 and 2 show a huge depression and recovery of the wind generator rotor speed, compared to scenario 3 with mitigated effects of the grid fault on the rotor speed during transient state. This is because an increase in the turn-on resistance of the IGBTs of the PMSG wind turbine power converters would decrease the circulation of currents and boost the capability of the switching mode of the converter's legs during transient state. Hence, this topology could be a cheaper way to enhance the FRT of the PMSG VSWT during transient condition.

4.5.2 EFFECTS OF THE INSULATED GATE BIPOLAR EXCITATION PARAMETERS ON THE PMSG CONSIDERING OVERVOLTAGE PROTECTION SCHEME (OVPS)

A further investigation was carried out for the three scenarios, using the same conditions in Section 4.5.1, of this chapter, considering the use of DC-chopper braking resistor for 110% overvoltage protection, as shown in Table 4.1. The PMSG variables for this case are shown in Figures 4.8–4.12.

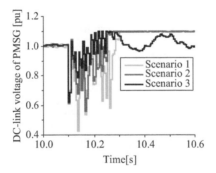

FIGURE 4.8 DC-link voltage of PMSG with OVPS.

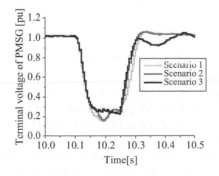

FIGURE 4.9 Terminal voltage of PMSG with OVPS.

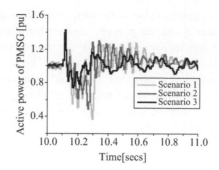

FIGURE 4.10   Active power of PMSG with OVPS.

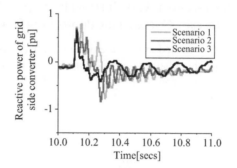

FIGURE 4.11   Reactive power of PMSG with OVPS.

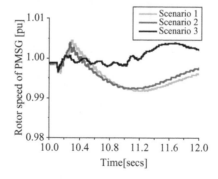

FIGURE 4.12   Rotor speed of PMSG with OVPS.

From Figure 4.8, with the help of the DC-chopper connected at the GSC of the PMSG with turbine, the DC-link voltage was maintained within the permissible limit of 110% as set in Table 4.1, for all the scenarios. It could be observed from Figure 4.8 that the DC-link OVPS was able to mitigate the effects of the grid disturbance, oscillations, overshoot and undershoot, for scenario 3, which is usually caused by grid fault, due to the fragile nature of the power converters of the wind generator. Thus, the vulnerable power converters of the wind generator are protected during transient

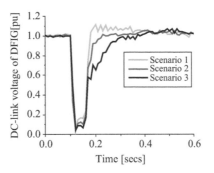

FIGURE 4.13 DC-link voltage of DFIG.

state. The response of the terminal voltage of the wind generator's variable in Figure 4.9 indicates that the OVPS scheme could help reduce the voltage dip and overshoot in the terminal voltage of the wind turbine for scenario 3. A smoother active power of the wind generator was achieved for scenario 3 in Figure 4.10, because the effective control of the DC-link voltage by the OVPS has a direct effect on the active power of the wind turbine. The reactive power dissipation was less for scenario 3 with the help of the OVPS, as shown in Figure 4.11. In Figure 4.12, the rotor speed of the PMSG wind turbine was further improved, considering the OVPS for scenario 3. An improved performance of the variables of the PMSG wind turbine during transient state, in scenarios 1 and 2, respectively, was achieved using the OVPS for all the scenarios compared to the case without OVPS in Section 4.5.1 simulation results of this chapter, however, the OVPS has little or no effect on scenario 3, in the transient performance of the PMSG wind turbine. Thus, a cheaper and cost-effective way for the enhancement of the PMSG wind turbine is scenario 3, without the circuitry of DC-chopper braking resistor.

4.6 ANALYSIS OF THE DFIG WIND TURBINE CONSIDERING THE POWER CONVERTERS EXCITATION PARAMETERS

In order to compare the performance of the effects of the power converters on the transient stability of the PMSG wind turbine, a further analysis was carried out in this section to investigate the effects of the power converter on the DFIG wind turbine. The same condition of operation was considered as above for the PMSG wind turbine. The simulation results for some of the key variables of the DFIG are shown in Figures 4.13–4.17, as presented in Chapter 3 of this book.

Figure 4.13 shows that for scenario 1 the DFIG terminal voltage experienced overshoot, while for scenario 2, the terminal voltage was not recovered on time. The effective performance of the DFIG wind turbine terminal voltage was observed in scenario 2. When compared to the PMSG wind turbine, scenario 3 was better than the other two scenarios. The reactive power and active power responses in Figures 4.14 and 4.15 show that the DFIG was effectively stable for scenario 2, compared to the other scenarios; however, scenario 3 gave the effective performance of the PMSG wind turbine. Though scenario 1 gave a better response for the DFIG in Figures 4.16

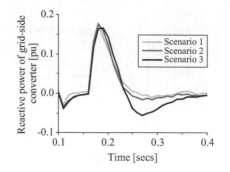

FIGURE 4.14   Reactive power of DFIG grid side converter

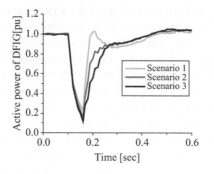

FIGURE 4.15   Active power of DFIG.

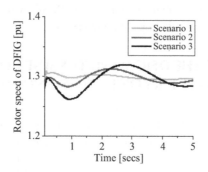

FIGURE 4.16   Rotor speed of DFIG.

and 4.17, for the rotor speed and terminal voltage variables, however, for the PMSG wind turbine, scenario 3 gave the most effective performance.

4.7   CHAPTER CONCLUSION

This chapter investigated the effects of the Insulated Gate Bipolar Transistor of the PMSG wind turbine power converters, during transient state. A severe three-phase to

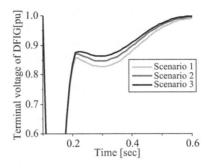

FIGURE 4.17 Terminal voltage of DFIG.

ground-bolted fault was considered in this study during transient condition, in order to test the robustness and rigidity of the controllers of the wind generator. Various scenarios were considered, where the turn-on resistance of the IGBTs varied, while keeping the turn-off resistance, forward break over voltage and reverse withstand voltage constant. This chapter also considered the performance of the wind generator with and without the use of DC-chopper overvoltage protection scheme, connected to the grid side converter of the wind generator. The presented results show that a high turn-on resistance of the IGBTs of the PMSG power converters could help improve the performance of the variables of the wind generator during transient state, even without considering overvoltage protection scheme. This is because an increase in the turn-on resistance would limit the excess circulating current that normally occurs during transient state of the wind turbine. More so, the performance of the PMSG wind turbine seems to be better with a slightly higher turn-on resistance compared to the turn-on resistance value for the DFIG wind turbine, based on the presented results. This could be as a result of the full-back-back power converter rating for the PMSG wind turbine, as compared to only 20–30% power converter rating for the DFIG wind turbine. As part of future scope, an optimization approach using Dragon Fly Algorithm, is the next phase of this work, for optimal performance of the wind generator.

REFERENCES

[1] M. M. R. Singaravel, and S. A. Daniel, "MPPT with single DC-DC converter and inverter for grid connected hybrid wind-driven PMSG-PV system," *IEEE Transactions on Industrial Electronics*, vol. 62, pp. 4849–4857, 2015.
[2] A. S. Ahmed, A. Noura, A. Nour, M. A. Ahmed, and S. E. A. Walid, "Fuzzy logic-based MPPT technique for PMSG wind generation system," *International Journal of Renewable Energy Research*, vol. 9, no. 4, pp. 1751–1760, 2019.
[3] K. E. Okedu, S. M. Muyeen, R. Takahashi, and J. Tamura, "Protection schemes for DFIG considering rotor current and DC-link voltage," *24th IEEE-ICEMS (International Conference on Electrical Machines and System)*, Beijing, China, August 2011, pp. 1–6.
[4] M. Rosyadi, S. M. Muyeen, R. Takahashi, and J. Tamura, "A design fuzzy logic controller for a permanent magnet wind generator to enhance the dynamic stability of wind farms," *Applied Sciences*, vol. 2, pp. 780–800, 2012, doi: 10.3390/app2040780.

[5] G. Michalke, A. Hansen, and T. Hartkopf, "Control strategy of a variable speed wind turbine with multi-pole permanent magnet synchronous generator," *Proceedings of European Wind Energy Conference and Exhibition*, Milan, Italy, 7–10 May 2007.

[6] M. Nasiri, and R. Mohammadi, "Peak current limitation for grid-side inverter by limited active power in PMSG-based wind turbines during different grid faults," *IEEE Transactions on Sustainable Energy*, vol. 8, pp. 3–12, 2017.

[7] A. Gencer, "Analysis and control of fault ride through capability improvement PMSG based on WECS using active crowbar system during different fault conditions," *Elektronika ir Elektrotechnika*, vol. 24, pp. 64–69, 2018.

[8] D. M. Yehia, D. A. Mansour, and W. Yuan, "Fault ride-through enhancement of PMSG wind turbines with DC microgrids using resistive-type SFCL," *IEEE Transactions on Applied Superconductivity*, vol. 28, pp. 1–5, 2018.

[9] X. Zeng, J. Yao, Z. Chen, W. Hu, Z. Chen, and T. Zhou, "Co-ordinated control strategy for hybrid wind farms with PMSG and FSIG under unbalanced grid voltage condition," *IEEE Transactions on Sustainable Energy*, vol. 7, pp. 1100–1110, 2016.

[10] G. Ghatikar, S. Mashayekh, M. Stadler, R. Yin, and Z. Liu, "Distributed energy systems integration and demand optimization for autonomous operations and electric grid trans-actions," *Applied Energy*, vol. 167, pp. 432–448, 2016.

[11] S. Alepuz, A. Calle, M. S. Busquets, S. Kouro, and B. Wu, "Use of stored energy in PMSG rotor inertia for low-voltage ride-through in back-to-back NPC converter-based wind power systems," *IEEE Transactions on Industrial Electronics*, vol. 60, pp. 1787–1796, 2013.

[12] M. E. Hossain, "A non-linear controller based new bridge type fault current limiter for transient stability enhancement of DFIG based wind farm," *Electric Power Systems Research*, vol. 152, pp. 466–484, 2017.

[13] A. B. Pulido, J. Romero, and H. C. Enriquez, "Robust active disturbance rejection control for LVRT capability enhancement of DFIG-based wind turbines," *Control Engineering Practice*, vol. 77, pp. 174–189, 2018.

[14] M. Youjie, T. Long, Z. Xuesong, L. Wei, and S. Xueqi, "Analysis and control of wind power grid integration based on a permanent magnet synchronous generator using a fuzzy logic system with linear extended state observer," *Energies* vol. 12, pp. 2862, 2019, doi: 10.3390/en12152862.

[15] A. S. Mahmoud, M. H. Hany, Z. A. Haitham, E. E. El-Kholy, and A. M. Sabry, "An adaptive fuzzy logic control strategy for performance enhancement of a grid-connected PMSG-Based wind turbine," *IEEE Transactions on Industrial Informatics*, vol. 15, no. 6, pp. 3163–3173, 2019.

[16] A. R. A. Jerin, P. Kaliannan, and U. Subramaniam, "Improved fault ride through capability of DFIG based wind turbines using synchronous reference frame control based dynamic voltage restorer," *ISA Transactions*, vol. 70, pp. 465–474, 2017.

[17] L. Chen, H. He, H. Chen, L. Wang, L. Zhu, Z. Shu, J. Yang, and F. Tang, "Study of a modified flux-coupling-type SFCL for efficient fault ride-through in a PMSG wind turbine under different types of faults," *Journal of Electrical and Computer Engineering*, vol. 40, no. 3, pp. 189–200, 2017.

[18] A. Moghadasi, A. Sarwat, and J. P. M. Guerrero, "Multiobjective optimization in combinatorial wind farms system integration and resistive SFCL using analytical hierarchy process," *Renewale Energy*, vol. 94, pp. 366–382, 2016.

[19] C. Huang, X. Y. Xiao, Z. Zheng, and Y. Wang, "Cooperative control of SFCL and SMES for protecting PMSG-based WTGs under grid faults," *IEEE Transactions on Applied Superconductivity*, vol. 29, no. 2, pp. 1–6, 2019.

[20] M. Firouzi, M. Nasiri, M. Benbouzid, and G. B. Gharehpetian, "Application of multi-step bridge-type fault current limiter for fault ride-through capability enhancement of permanent magnet synchronous generator-based wind turbines," *International Transactions on Electrical Energy*, vol. 30, no. 11, p. e12611, 2020.

[21] K. E. Okedu, and H. Barghash, "Enhancing the performance of DFIG wind turbines considering excitation parameters of the insulated gate bipolar transistors and a new PLL scheme," *Frontiers in Energy*, vol. 8, p. 373, 2021.

[22] K. E. Okedu, and H. Barghash, "Enhancing the transient state performance of permanent magnet synchronous generator based variable speed wind turbines using power converters excitation parameters," *Frontiers in Energy Research-Smart Grids*, vol. 9, pp. 109–120, Article 655051, 2021, doi: 10.3389/fenrg.2021.655051.

[23] Toshiba Electronic Devices & Storage Corporation, IGBTs (Insulated Gate Bipolar Transistor), Application note,, pp. 6–8, 2018–2022, Minato-ku, Tokyo, Japan. https://toshiba.semicon-storage.com/ap-en/top.html

[24] PSCAD/EMTDC Manual, Version 4.6.0; Manitoba HVDC Lab.: Winnipeg, MB, Canada, 2016.

5   DFIG and PMSG Machine Parameters

5.1   CHAPTER INTRODUCTION

Variable Speed Wind Turbines (VSWTs) are the new norms for the installation of wind farms. These classes of wind turbines have high efficiency in capturing energy, with effective voltage control [1]. The Doubly Fed Induction Generator (DFIG) and Permanent Magnetic Synchronous Generator (PMSG) with back-to-back power converter type technologies have become the two wind generator alternatives, commonly employed as VSWTs. The former has a gearbox, and only 20–30% of the generator rating is required for its operating speed range of 0.7–1.3 per unit (p.u). The latter's drawback is high cost due to its full-rated power converters.

In the DFIG wind turbine structure, the back-to-back power converter is in between the rotor or machine side and the stator or grid side. Based on the available wind speeds, this type of wind turbine can operate at a wide range, for better wind energy capture [2, 3]. In the DFIG wind turbine, rebuilding of the terminal voltage after grid disturbance is much easier due to the pitch and dynamic slip control strategy [4, 5]. More so, the control of active and reactive power through decoupling principles is much easier in this class of wind turbine. The power converters of the DFIG wind turbine go into standby mode at lower voltages [6–9]; however, during grid fault, above the threshold voltages, there is fast synchronization of this type of wind generator to the power grid.

On the other hand, the alternative to the DFIG wind turbine in wind energy conversion is the PMSG wind turbine technology. The PMSG wind turbine has a full-rated back-to-back power converter tied to the power grid. Consequently, there is maximum flexibility in this type of wind generator, compared to the DFIG topology [10, 11]. Thus, real and reactive power control are more effective when using the PMSG wind turbine. However, high initial cost is the major shortcoming of this class of wind turbine system [12].

Based on the literature, there are many Fault Ride Through (FRT) control strategies in the enhancement of DFIG and PMSG wind turbines. Fault current limiters [13–15], crowbar switch and DC chopper circuitry [16, 17] and sliding mode controls [18, 19] are some of the FRT schemes already reported in the literature for the improvement of the transient stability of either or both wind turbines. The assessment of DFIG using various control methods was carried out in [20–22], with emphasis on the use of Maximum Power Point Tracking (MPPT) pitch angle controller, considering different algorithms in [23], while peak current limitation and MPPT were employed in [24, 25], respectively. In [26], a series fault current limiter was employed with metal oxide varistor, to improve the Low Voltage Ride Through

DOI: 10.1201/9781003350910-5

(LVRT) of DFIG-based wind farms. The application of multi-step bridge-type fault current limiter for the PMSG wind turbine was reported in [27], for the improved performance of the wind turbine FRT performance. A neuro fuzzy logic controlled parallel resonance type fault current limiter scheme was employed in [28], to enhance the FRT capability of a wind farm composed of DFIG wind turbines, while an entire power systems using fuzzy logic controlled capacitive-bridge-type fault current limiter scheme was investigated in [29]. A further approach of using dynamic multi-cell fault current limiter was reported in [30], to improve the FRT performance of wind farms based on DFIG control, while a sliding mode controller based on bridge-type fault current limiter was used for the DFIG FRT improved performance in [31].

Machine parameters help in understanding the performance of wind generators. These parameters are additional quantities that influence the behavior of the wind turbines. However, compared to input variables, the machine parameters are held constant or changes slowly in the operation of the wind turbines [32]. Machine parameters present some unique functions that should be observed in order to improve the maximum power generation output and efficiency of the wind turbines [33]. In [34], the evaluation of the mechanical parameters affecting wind turbine power generation was carried out, however, the parameters investigated were limited to the swept area, air density, wind speed and power coefficient of the wind turbine. In [35], wind turbine parameters were estimated and these effects were used to quantify their dynamic behavior. The study was limited to the estimation of certain parameters, without extensive study of their effects during transient conditions. In [36], an optimized parameter design for the blades of wind turbine considering Schmitz theories and aerodynamics forces was reported.

In light of the above, this chapter presents the effects of the electrical parameters of wind turbines during grid fault. There are limited reports on the effects of electrical parameters considering the DFIG and PMSG VSWTs in the literature. Therefore, this chapter tends to bridge this research gap. The DFIG and PMSG wind turbine characteristics, modeling and control strategies were presented already in Chapter 1 of this book. The extensive investigation of the effects of the various parameters were considered, by varying some machine parameters while keeping the other machine parameters constant. In the DFIG analysis, nine scenarios were considered, while in the PMSG analysis, ten scenarios were considered. The machine parameters considered for the DFIG wind turbine are magnetizing inductance, stator leakage inductance, wound rotor leakage inductance, stator resistance, wound rotor resistance and the angular moment of inertia. The machine parameters considered for the PMSG wind turbine are Stator winding resistance, Stator leakage reactance, D: Unsaturated reactance, Q: Unsaturated reactance, D: Damper winding resistance, D: Damper winding reactance, Q: Damper winding reactance, Q: Damper winding resistance and magnetic strength. The same wind turbine capacity and the same fault condition of operation were considered for both wind turbine technologies, and the turbines were operated at their rated speed based on their MPPT wind turbine characteristics. The simulation analysis was carried out using Power System Computer Aided Design and Electromagnetic Transient Including DC (PSCAD/EMTDC) [37].

5.2 TURBINE CHARACTERISTICS AND MODEL OF DFIG WIND TURBINE

The DFIG wind turbine characteristics already exist in Chapter 1 of this book. In Figure 5.1, the total power is $(P_T = P_s + P_g)$, and it is delivered based on the available wind speed through the rotor and stator circuits of the wind turbine to the PCC, for $\omega_r > \omega_s$ (rotor speed greater than synchronous speed). And only active power P_s is delivered to the PCC through the stator circuitry $(\omega_r < \omega_s)$, while the absorption of the active power P_r from the PCC is done via the power converter [38]. In the d-q orientation stator frame synchronous axis, the application of vector control would lead to the wind generator's model on the 5th-order d-q representation [39]. Therefore, the stator, rotor and the d-q components considering Figure 5.1 equivalent circuit for the DFIG wind turbine are:

$$
\begin{bmatrix} \psi_{ds} \\ \psi_{qr} \\ \psi_{dr} \\ \psi_{qr} \end{bmatrix} = \begin{bmatrix} L_s & 0 & L_m & 0 \\ 0 & L_s & 0 & L_m \\ L_m & 0 & L_r & 0 \\ 0 & L_m & 0 & L_r \end{bmatrix} \begin{bmatrix} I_{ds} \\ I_{qr} \\ I_{dr} \\ I_{qr} \end{bmatrix}
\tag{5.1}
$$

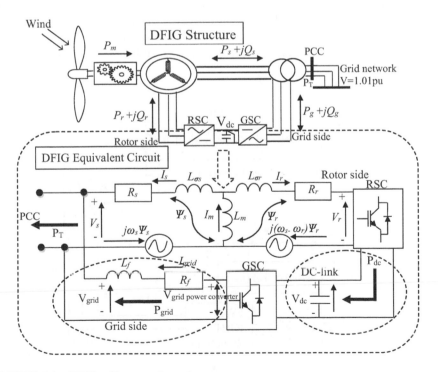

FIGURE 5.1 DFIG grid connection scheme.

$$
\begin{bmatrix} V_{ds} \\ V_{qs} \end{bmatrix} = \begin{bmatrix} R_s & 0 \\ 0 & R_s \end{bmatrix} \begin{bmatrix} I_{ds} \\ I_{qs} \end{bmatrix} + \frac{d}{dt} \begin{bmatrix} \psi_{ds} \\ \psi_{qs} \end{bmatrix} + \begin{bmatrix} 0 & -\omega_s \\ \omega_s & 0 \end{bmatrix} \begin{bmatrix} \psi_{ds} \\ \psi_{qs} \end{bmatrix}
\tag{5.2}
$$

$$
\begin{bmatrix} V_{dr} \\ V_{qr} \end{bmatrix} = \begin{bmatrix} R_r & 0 \\ 0 & R_r \end{bmatrix} \begin{bmatrix} I_{dr} \\ I_{qr} \end{bmatrix} + \frac{d}{dt} \begin{bmatrix} \psi_{dr} \\ \psi_{qr} \end{bmatrix} + \begin{bmatrix} 0 & -\omega_{slip} \\ \omega_{slip} & R_s \end{bmatrix} \begin{bmatrix} \psi_{dr} \\ \psi_{qr} \end{bmatrix}
\tag{5.3}
$$

From Equations (5.1–5.3), the direct and quadrature axis stator reference frame components are given by d and q, respectively, while s and r, m are the stator, rotor and mutual quantities and component. V, I and ψ represent voltage, current and flux vector quantities. L and R are the inductance and resistance. $\omega_{slip} = (\omega_s - \omega_r)$, ω_r and ω_s are the slip, electrical rotor and synchronous speeds, respectively. L_r and L_s are the total inductance in rotor and stator circuits given by $L_s = L_m + L_{\sigma s}$ and $L_r = L_m + L_{\sigma r}$.

The DC-link dynamics with respect to the DC power is given as:

$$
P_{dc} = C_{dc} V_{dc} \frac{dV_{dc}}{dt} = P_g - P_r
\tag{5.4}
$$

In Equation (5.4), C_{dc}, V_{dc}, P_{dc} are the DC-link capacitor, DC-link voltage and DC-link power. The DFIG Rotor Side Converter (RSC) control scheme is shown in Figure 5.2, where the q- and d-axis rotor currents i_{qr} and i_{dr} regulate the active and reactive power (P_s, Q_s), of the stator, respectively. The stator power of the wind generator is obtained by the MPPT scheme of the wind turbine system. The Grid Side

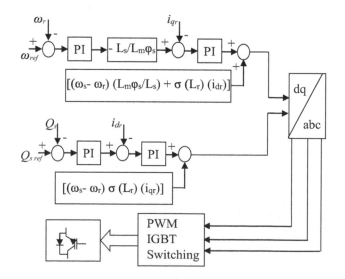

FIGURE 5.2   The rotor side converter control circuit of the DFIG.

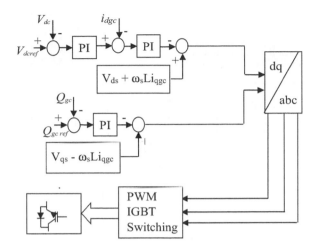

FIGURE 5.3   The grid side converter control circuit of the DFIG.

Converter (GSC) of the wind generator in Figure 5.3, utilizes the ac grid reference frame to regulate the DC-link voltage and flow of reactive power exchange (absorption and dissipation) in the PCC according to the power flow direction of the rotor circuit of the wind turbine. The dq-to-abc and abc-to-dq voltage angles transformation based on the ac voltage synchronization is achieved with the help of Phase Lock Loop (PLL) control strategy.

5.3   TURBINE CHARACTERISTICS AND MODEL OF PMSG WIND TURBINE

The PMSG wind turbine characteristics already exist in Chapter 1 of this book. Figure 5.4 shows the controller of the Machine Side Converter (MSC) and the GSC of the PMSG wind turbine. The MSC regulates both active and reactive power variables of the wind generator. The angle position rotor (θ_r), helps in achieving the abc to dq transformation by the rotor speed of wind generator. The d-axis current (I_{sd}) controls and the active power (P_s) of the wind generator, while the q-axis currents (I_{sq}) regulates the reactive power (Q_s) of the wind generator, respectively. The MPPT technique for the characteristic is employed in the reference active power (P_{ref}). Usually, the reference reactive power (Q_s^*) is fixed at 0, to obtain effective operation of power factor 1. Three-phase reference voltages (V_{sa}^*, V_{sb}^*, V_{sc}^*) are generated for the PWM switching, through the outputs of the current controller, based on the voltage references V_{sd}^* and V_{sq}^*. The GSC control of the wind generator is also shown in Figure 5.4 and is regulated considering the d-q rotating reference frame, based on the voltage of the grid as the same as the speed of rotation. The Park transformation is used in converting (I_{ga}, I_{gb}, I_{gc}) and (V_{ga}, V_{gb}, V_{gc}) three-phase voltage and current of the grid into their rotating reference d-q frame. The extraction of the phase angle (θ_g) of the grid side is done by considering the PLL [40].

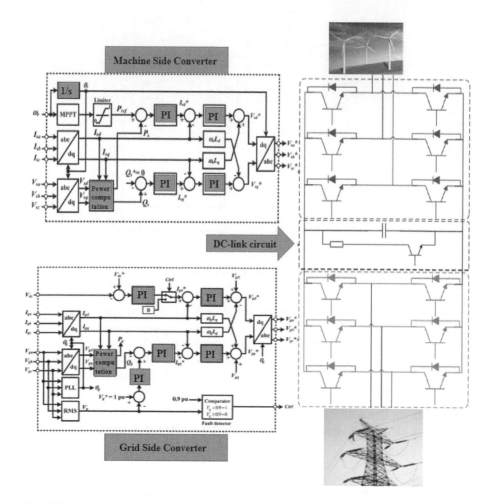

FIGURE 5.4   Control strategy of the PMSG wind turbine.

5.4   EVALUATION OF THE SYSTEM PERFORMANCE

5.4.1   DFIG Wind Turbine Machine Parameters' Evaluation

The simulation was carried out considering the nine scenarios summarized in Table 5.1 [41], for the DFIG machine parameters in per unit values. The fault type is a severe three-phase of 100 ms happening at 0.1 s, with the circuit breakers operation sequence opening and reclosing at 0.2 s and 1.0 s, respectively, on the faulted line at the terminals of the DFIG wind turbine. The system performance was evaluated using PSCAD/EMTDC environment. Some of the simulation results for the cases considered are shown in Figures 5.5–5.14, based on the design machine parameters obtained from the manufacturers of the wind turbine.

In the first part of the analysis, cases 1–3 were considered. Case 1 machine parameters values were the original values used by the DFIG wind generator. The machine

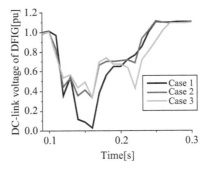

FIGURE 5.5 DC-link voltage of DFIG (cases 1–3).

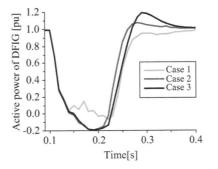

FIGURE 5.6 Active power of DFIG (cases 1–3).

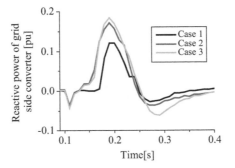

FIGURE 5.7 Reactive power of GSC of DFIG (cases 1–3).

parameters were varied for cases 2 and 3, considering a uniform increment for all the parameters. In Figure 5.5, when the machine parameters were too high or too low, the DC-link voltage of the DFIG wind turbine response was not improved, compared to case 2. In cases 1 and 3, the DC-link voltage variable had more voltage dip with less recovery time. Because the DC-link voltage is related to the response of the active power of the wind generator, the same pattern of response could be observed in

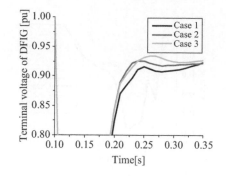

FIGURE 5.8   Terminal voltage of DFIG (cases 1–3).

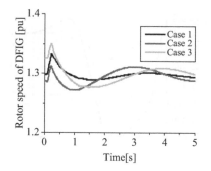

FIGURE 5.9   Rotor speed of DFIG (cases 1–3).

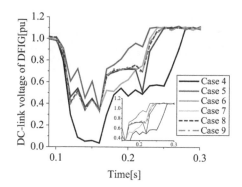

FIGURE 5.10   DC-link voltage of DFIG (cases 4–9).

Figure 5.6 for the active power of the DFIG wind generator, where case 2 gave an improved performance. In Figure 5.7, more reactive power was dissipated by increasing the machine parameters of the DFIG wind generator; consequently, an improved terminal voltage response was observed in Figure 5.8. However, in case 3, there was more overshoot of the terminal voltage variable compared to case 2. Figure 5.9 shows

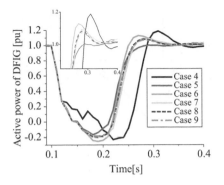

FIGURE 5.11 Active power of DFIG (cases 4–9).

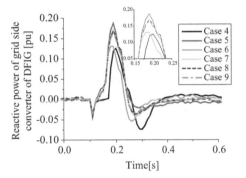

FIGURE 5.12 Reactive power of GSC of DFIG (cases 4–9).

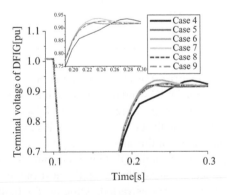

FIGURE 5.13 Terminal voltage of DFIG (cases 4–9).

that lower machine parameters of the DFIG wind generator would result in less oscillation of the rotor speed variable during transient state; however, it could be trade-off compared to the improved performance obtained for the other variables of the wind generator.

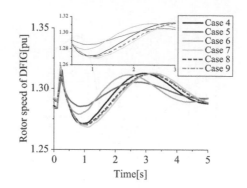

FIGURE 5.14 Rotor speed of DFIG (cases 4–9).

TABLE 5.1
DFIG Machine Parameters [41]

DFIG wind turbine (5 MVA)	Magnetizing inductance (pu)	Stator leakage inductance (pu)	Wound rotor leakage inductance (pu)	Stator resistance (pu)	Wound rotor resistance (pu)	Angular moment of inertia (s)
Case 1	3.5	0.15	0.15	0.01	0.01	3.0
Case 2	3.9	0.19	0.19	0.05	0.05	3.5
Case 3	4.3	0.23	0.23	0.09	0.09	3.9
Case 4	3.9	0.19	0.19	0.01	0.05	3.5
Case 5	3.9	0.19	0.19	0.05	0.01	3.5
Case 6	3.9	0.19	0.19	0.05	0.05	3.0
Case 7	3.5	0.19	0.19	0.05	0.05	3.5
Case 8	3.9	0.15	0.19	0.05	0.05	3.5
Case 9	3.9	0.19	0.15	0.05	0.05	3.5

A further analysis was carried out considering the best values for case 2, where improved performance of the DFIG wind turbine variables were obtained during transient state. The machine parameters were varied and kept constant for cases 4–9 in Table 5.1, in order to understand the behavior of the DFIG variables during grid fault. Some of the simulation results for this analysis are given in Figures 5.10–5.14.

In case 4, the stator resistance of the DFIG wind turbine was varied, while keeping the other machine parameters constant based on the best-obtained values in case 2. In case 5, the wound rotor resistance was varied, while in case 6, the angular moment of inertia was varied. In cases 7 and 8, the magnetizing inductance and the stator leakage inductance were varied, respectively. In case 9, the wound rotor leakage inductance was varied. It could be observed that, varying the stator resistance has a huge effect on the DC-link voltage and active power of the DFIG in Figures 5.10 and 5.11. Also, cases 5 and 6 have significance effect of the DC-link voltage of the wind generator. However, cases 7–9 have no effect on the performance of the DFIG DC-link voltage during transient state. In Figure 5.12, there is much effect on the reactive power of the DFIG wind turbine, considering cases 4 and 6. However, more reactive power was dissipated for case 5, while cases 7–9 have no significant effect on the reactive power of the wind turbine. Cases 4 and 6 have more effect on the terminal voltage of the DFIG wind turbine, while cases 5–9 have no significant effect on the terminal voltage variable of the wind turbine. Case 5 gave the best performance of the rotor speed of the DFIG wind turbine, compared to the other cases. It can be seen from the above results that the variation of the stator resistance in case 4 has much influence on the transient stability of the DFIG during grid fault.

5.4.2   PMSG Wind Turbine Machine Parameters' Evaluation

Simulation was carried out considering the ten scenarios summarized in Table 5.2, for the PMSG machine parameters in per unit values. The same fault conditions and scenarios used for the DFIG wind turbine in Section 5.4.1, of the simulation results in this chapter, were considered for the PMSG wind turbine analysis in this section. The fault type is a severe three-phase of 100 ms happening at 10.1 s, with the circuit breakers operation sequence opening and reclosing at 10.2 s and 11.0 s, respectively, on the faulted line at the terminals of the PMSG wind turbine. In the first part of the analysis, cases 1–3, were considered. Case 1 machine parameters values were the original values used by the PMSG wind generator. The machine parameters were varied for cases 2 and 3, considering a uniform increment for all the parameters. Some of the simulation results are shown in Figures 5.15–5.24, based on the design machine parameters obtained from the manufacturers of the wind turbine.

In Figures 5.15 and 5.16, case 3 gave the best performance of the DC-link and active power of the PMSG wind turbine during transient, with faster recovery, short settling time and less oscillations. The same behavior could be observed for the reactive power of the GSC, terminal voltage and rotor speed of the PMSG wind turbine in Figures 5.17–5.19, respectively. From the results, a too low value of the machine parameters of the PMSG wind turbine led to poor performance, compared to the use of higher values of the PMSG machine parameters. Thus, the best values of case 3 were used for further analysis of the PMSG wind turbine machine parameters in cases 4–10 by varying the parameters. In case 4, the Q: Damper winding reactance was varied and other machine parameters were kept constant, while in case 5, the D: Damper winding reactance was varied. In case 6, the D: Damper winding resistance was varied, while in case 7, Q: Unsaturated reactance was varied. The D: Unsaturated reactance was varied in case 8, while in case 9, the stator leakage reactance was varied. Case 10 considered the variation of the stator winding resistance machine param eters of the PMSG wind turbine. A further analysis of the effect of the machine parameters was considered by varying these parameters for cases 4–10 as summarized in Table 5.2. Some of the simulation results are shown in Figures 5.20–5.24.

Cases 8 and 10 have huge influence on the DC-link voltage and active power of the PMSG wind turbine as shown in Figures 5.20 and 5.21, while cases 4–7 and 9 have little or no influence on the PMSG variables. In Figure 5.22, the reactive power variable of the PMSG is much affected in case 10, compared to the other cases, while in Figure 5.23, cases 5 and 10 have a little influence on the terminal voltage variable

TABLE 5.2
PMSG Machine Parameters

PMSG wind turbine (5 MVA)	Stator winding resistance (pu)	Stator leakage reactance (pu)	D: Unsaturated reactance (pu)	Q: Unsaturated reactance (pu)	D: Damper winding resistance (pu)	D: Damper winding reactance (pu)	Q: Damper winding reactance (pu)	Q: Damper winding resistance (pu)	Magnetic strength (pu)
Case 1	0.01	0.064	1.00	0.70	0.055	1.00	0.183	1.175	1.4
Case 2	0.05	0.068	1.04	0.74	0.059	1.04	0.187	1.179	1.44
Case 3	0.09	0.072	1.08	0.78	0.063	1.08	0.191	1.183	1.48
Case 4	0.09	0.072	1.08	0.78	0.063	1.08	0.183	1.175	1.48
Case 5	0.09	0.072	1.08	0.78	0.063	1.00	0.191	1.183	1.48
Case 6	0.09	0.072	1.08	0.78	0.055	1.08	0.191	1.183	1.48
Case 7	0.09	0.072	1.08	0.70	0.063	1.08	0.191	1.183	1.48
Case 8	0.09	0.072	1.00	0.78	0.063	1.08	0.191	1.183	1.48
Case 9	0.09	0.064	1.08	0.78	0.063	1.08	0.191	1.183	1.48
Case 10	0.01	0.072	1.08	0.78	0.063	1.08	0.191	1.183	1.48

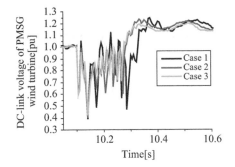

FIGURE 5.15 DC-link voltage of PMSG (cases 1–3).

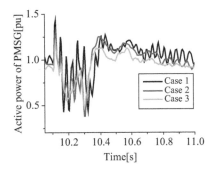

FIGURE 5.16 Active power of PMSG (cases 1–3).

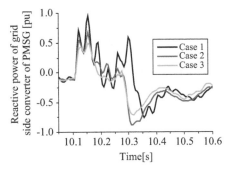

FIGURE 5.17 Reactive power of GSC of PMSG (cases 1–3).

of the PMSG, compared to the other cases with no significant effect. Figure 5.24 shows that case 10 has huge influence on the rotor speed of the PMSG wind turbine, by increasing its value compared to case 5, with less effect and the other cases with insignificant effect on the variable of the PMSG. From the results, it could be observed that variation in the stator winding resistance of the PMSG wind turbine machine parameter in case 10 has great significance in its performance during grid fault.

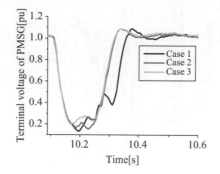

FIGURE 5.18   Terminal voltage of PMSG (cases 1–3).

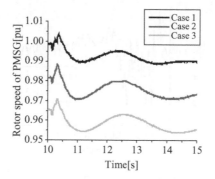

FIGURE 5.19   Rotor speed of PMSG (cases 1–3).

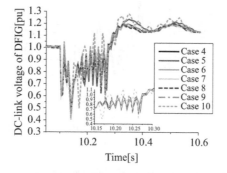

FIGURE 5.20   DC-link voltage of PMSG (cases 4–10)

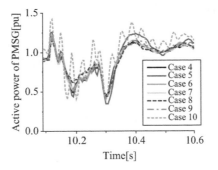

FIGURE 5.21 Active power of PMSG (cases 4–10).

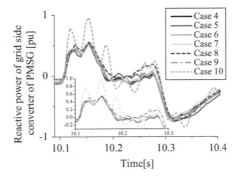

FIGURE 5.22 Reactive power of GSC of PMSG (cases 4–10).

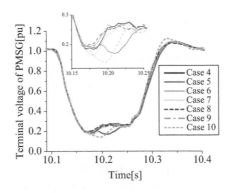

FIGURE 5.23 Terminal voltage of PMSG (cases 4–10).

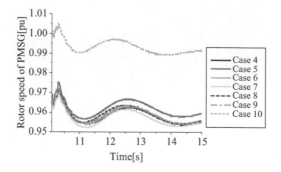

FIGURE 5.24   Rotor speed of PMSG (cases 4–10).

5.5   CHAPTER CONCLUSION

The performance of two VSWTs, DFIG and (PMSG) machine parameters, were investigated during severe grid fault, considering their machine parameters. Using too low or too high machine parameter values, affect the performance of the DFIG wind turbine during transient state, with reference to the commercial obtained values used for both wind turbines (case 1) in this study, from Hitachi Laboratory, Hokkaido, Japan. Moderate values of machine parameters of the DFIG wind turbine are better for improved performance during grid fault. Also, the stator resistance parameter has much influence on the variables of the DFIG wind turbines, during transient state. Too low machine parameters affect the PMSG wind turbines during transient state. The stator winding resistance of the PMSG affects greatly the performance of the PMSG variables during transient state.

Therefore, for both wind turbine technologies, the stator winding resistance machine parameter has a significant effect on the wind turbine, compared to the other parameters. Consequently, a careful selection of the stator winding resistance based on the design of the wind turbine could help improve the FRT challenges posed by these wind turbines during transient state. This may be a cost-effective way without considering external circuitry when compared to most existing FRT solutions for wind turbines.

REFERENCES

[1] K. E. Okedu, "Introductory," In Power System Stability, UK: INTECH, 2019, pp. 1–10.
[2] S. Bozhko, G. Asher, R. Li, J. Clare, and L. Yao, "Large offshore DFIG-based wind farm with line-commutated HVDC connection to the main grid: Engineering studies," *IEEE Transactions on Energy Conversion*, vol. 23, no. 1, 2008.
[3] M. Godoy Simoes, and F. A. Farret, *Renewable Energy Systems: Design and Analysis with Induction Generators*, Boca Raton, FL: CRC Press, LLC, 2004.
[4] S. Santos, and H. T. Le, "Fundamental time-domain wind turbine models for wind power studies," *Renewable Energy*, vol. 32, pp. 2436–2452, 2007.
[5] M. Haberberger, and F. W. Fuchs, "Novel protection strategy for current interruptions in IGBT current source inverters," *Proceedings EPE-PEMC*, Oslo, Norway, 2004.

[6] K. E. Okedu, S. M. Muyeen, R. Takahashi, and J. Tamura, "Wind farms fault ride through using DFIG with new protection scheme," *IEEE Transactions on Sustainable Energy*, vol. 3 no. 2, pp. 242–254, 2012.

[7] A. A. El-Sattar, N. H. Saad, and M. Z. S. El-Dein, "Dynamic response of doubly fed induction generator variable speed wind turbine under fault," *Electric Power System Research*, vol. 78, pp. 1240–1246, 2008.

[8] K. E. Okedu, and H. Barghash, "Investigating variable speed wind turbine transient performance considering different inverter schemes and SDBR," *Frontiers in Energy Research-Smart Grids*, vol. 8, pp. 1–16, Article 604338, 2021.

[9] K. E. Okedu, and H. Barghash, "Enhancing the performance of DFIG wind turbines considering excitation parameters of the insulated gate bipolar transistors and a new PLL scheme," *Frontiers in Energy Research-Smart Grids*, vol. 8, pp. 1–14, Article 620277, 2021.

[10] K. E. Okedu, and H. Barghash, "Enhancing the transient state performance of permanent magnet synchronous generator based variable speed wind turbines using power converters excitation parameters," *Frontiers in Energy Research-Smart Grids*, vol. 9, pp. 109–xx, Article 655051, 2021.

[11] K. E. Okedu, "Wind turbine driven by permanent magnetic synchronous generator," *Pacific Journal of Science and Technology PJST*, vol. 12, no. 2, pp. 168–175, 2011.

[12] J. Baran, and J. Andrzej, "An MPPT control of a PMSG-based WECS with disturbance compensation and wind speed estimation," *Energies*, vol. 13, no. 6344, pp. 1–20, 2020.

[13] K. E. Okedu, S. M. Muyeen, R. Takahashi, and J. Tamura, "Use of supplementary rotor current control in DFIG to augment fault ride through of wind farm as per grid requirement," *37th Annual Conference of IEEE Industrial Electronics Society (IECON 2011)*, Melbourne, Australia, 7–10 November 2011.

[14] K. E. Okedu, S. M. Muyeen, R. Takahashi, and J. Tamura, "Improvement of fault ride through capability of wind farm using DFIG considering SDBR," *14th European Conference of Power Electronics EPE*, Birmingham, UK, August, 2011, pp. 1–10.

[15] K. E. Okedu, "Enhancing DFIG wind turbine during three-phase fault using parallel interleaved converters and dynamic resistor" *IET Renewable Power Generation*, vol. 10, no. 6, pp. 1211–1219, 2016.

[16] R. Takahashi, J. Tamura, M. Futami, M. Kimura, and K. Idle, "A new control method for wind energy conversion system using double fed synchronous generators," *IEEJ Transactions on Power and Energy*, vol. 126, no. 2, pp. 225–235, 2006.

[17] K. E. Okedu, S. M. Muyeen, R. Takahashi, and J. Tamura, "Comparative study between two protection schemes for DFIG-based wind generator," *23rd IEEE-ICEMS (International Conference on Electrical Machines and Systems)*, Seoul, South Korea, October 2010, pp. 62–67.

[18] K. E. Okedu, S. M. Muyeen, R. Takahashi, and J. Tamura, "Stabilization of wind farms by DFIG-based variable speed wind generators," *International Conference of Electrical Machines and Systems (ICEMS)*, Seoul, South Korea, 10–13 October 2010, pp. 464–469.

[19] Y. Bekakra, and D. B. Attous, "Sliding mode controls of active and reactive power of a DFIG with MPPT for variable speed wind energy conversion," *Australian Journal of Basic and Applied Sciences*, vol. 5, no. 12, pp. 2274–2286, 2011.

[20] D. M. Ali, K. Jemli, M. Jemli, and M. Gossa, "Doubly fed induction generator, with crowbar system under micro-interruptions fault," *International Journal on Electrical Engineering and Informatics*, vol. 2, no. 3, pp. 216–231, 2010.

[21] D. B. Suthar, "Wind energy integration for DFIG based wind turbine fault ride through", *Indian Journal of Applied Research*, vol. 4, no 5, pp. 216–220, 2014.

[22] M. T. Lamchich, and N. Lachguer Matlab, "Simulink as simulation tool for wind generation systems based on doubly fed induction machines," *MATLAB-A Fundamental Tool for Scientific Computing and Engineering Applications*, vol. 2, Chapter 7, INTECH Publishing, 2012, pp. 139–160.

[23] A. Noubrik, L. Chrifi-Alaoui, P. Bussy, and A. Benchaib, "Analysis and simulation of a 1.5MVA doubly fed wind power in matlab sim powersystems using crowbar during power systems disturbances," *IEEE-2011 International Conference on Communications, Computing and Control Applications (CCCA)*, Hammamet, Tunisia, 2011.

[24] M. Nasiri, and R. Mohammadi, "Peak current limitation for grid-side inverter by limited active power in PMSG-based wind turbines during different grid faults," *IEEE Transactions on Sustainable Energy*, vol. 8, pp. 3–12, 2017.

[25] A. Gencer, "Analysis and control of fault ride through capability improvement PMSG based on WECS using active crowbar system during different fault conditions," *Elektronika ir Elektrotechnika*, vol. 24, pp. 64–69, 2018.

[26] M. M. Moghimyan, M. Radmehr, and M. Firouzi, "Series resonance fault current limiter (SRFCL) with MOV for LVRT enhancement in DFIG-based wind farms," *Electric Power Components and System*, vol. 47, no. 19–20, pp. 1814–1825, 2020, doi: 10.1080/15325008.2020.1731873.

[27] M. Firouzi, M. Nasiri, M. Benbouzid, and G. B. Gharehpetian, "Application of multi-step bridge-type fault current limiter for fault ride-through capability enhancement of permanent magnet synchronous generator-based wind turbines," *International Transactions on Electrical Energy*, vol. 30, no. 11, p. e12611, 2020.

[28] M. R. Islam, J. Hasan, M. R. Rahman Shipon, M. A. H. Sadi, A. Abuhussein, and T. K. Roy, "Neuro fuzzy logic controlled parallel resonance type fault current limiter to improve the fault ride through capability of DFIG based wind farm," *IEEE Access*, vol. 8, pp. 115314–115334, 2020.

[29] M. A. H. Sadi, A. AbuHussein, and M. A. Shoeb, "Transient performance improvement of power systems using fuzzy logic controlled capacitive-bridge type fault current limiter," *IEEE Transactions on Power Systems*, vol. 36, no. 1, 2021.

[30] M. R. Shafiee, H. Shahbabaei Kartijkolaie, M. Firouzi, S. Mobayen, and A. Fekih, "A dynamic multi-cell FCL to improve the fault ride through capability of DFIG-based wind farms," *Energies*, vol. 13, no. 20, 2020.

[31] M. Firouzi, M. Nasiri, S. Mobayen, and G. B. Gharehpetian, "Sliding mode controller-based BFCL for fault ride-through performance enhancement of DFIG-based wind turbines," *Complexity*, vol. 2020, pp. 1–12, 2020.

[32] D. Q. Nykamp, "Function machine parameters," From Math Insight. http://mathinsight.org/function_machine_parameters, 2020.

[33] A. T. Abolude, and W. Zhou, "Assessment and performance evaluation of a wind turbine power output," *Energies*, vol. 11, no. 1992, pp. 1–15, 2018.

[34] Construction Review, "Major parameters that influence wind turbines power output," 2019, https://constructionreviewonline.com/renewables/major-parameters-that-influence-wind-turbines-power-output/ (last accessed on 16 March 2021).

[35] J. Rose, and I. A. Hiskens, "Estimating wind turbine parameters and quantifying their effects on dynamic behavior," *Power and Energy Society General Meeting - Conversion and Delivery of Electrical Energy in the 21st Century, 2008 IEEE*, Pittsburgh, PA, pp. 1–6.

[36] R. Balijepalli, V. P. Chandramohan, and K. Kirankumar, "Optimized design and performance parameters for wind turbine blades of a solar updraft tower (SUT) plant using theories of Schmitz and aerodynamics forces," *Sustainable Energy Technologies and Assessments*, vol. 30, pp. 192–200, 2018.

[37] PSCAD/EMTDC Manual, Version 4.6.0; Manitoba HVDC Lab.: Winnipeg, MB, Canada, 2016.

[38] I. Zubia, J. X. Ostolaza, A. Susperrgui, and J. J. Ugartemendia, "Multi-machine transient modeling of wind farms, an essential approach to the study of fault conditions in the distribution network," Applied Energy, vol. 89, no. 1, pp. 421–429, 2012.

[39] X. Kong, Z. Z. Xianggen, and M. Wen, "Study of fault current characteristics of the DFIG considering dynamic response of the RSC," *IEEE Transactions Energy Conversion*, vol. 2, no. 2, pp. 278–287, 2014.

[40] K. E. Okedu, M. Al-Tubi, and S. Alaraimi, "Comparative study of the effects of machine parameters on DFIG and PMSG variable speed wind turbines during grid fault," *Frontiers in Energy Research-Smart Grids*, vol. 9, pp. 174–187, Article 681443, 2021, doi: 10.3389/fenrg.2021.681443.

[41] Hitachi Heavy Equipment Construction, wind turbines for power and energy laboratory, Ibaraki Service Center, Japan, 2010, https://www.hitachi-power-solutions.com/en/company/renewable-energy/

6 PMSG in Different Grid Strengths

6.1 CHAPTER INTRODUCTION

With the recent proliferation and penetration of wind farms into existing power grids, it is of paramount importance to conduct numerous studies to counter grid disturbances based on operational grid codes. Lately, due to computer and power electronics technology, soft computing schemes are been applied in wind turbines. The four computing algorithms commonly used are the predictive technique, fuzzy logic controllers, artificial neural networks and genetic algorithm [1]. The fuzzy logic controller was employed in [2], with some merits over traditional controllers. However, the sensitivity level to variations in parameters, architecture and operation of the system is low. Some works in the literature combined the traditional Proportional Integral Derivative (PID) controllers with fuzzy logic [3–5] for better performance. Youjie et al. [1] used the linear observer extended state system to further enhance the PMSG wind turbine-based performance, considering fuzzy controller, while in [6], control based on fuzzy employing adaptive strategy was used for the PMSG.

In weak grids, the challenges of network stability were a result of wind energy penetration based on the literature [7–9]. The Voltage Source Inverter (VSI) based on Pulse Width Modulation (PWM) is employed widely in interfacing sources with regard to renewable energy and the grid [10, 11]. The utilization of these inverters would cause stability issues in the power grid. The studies carried out in the literature show that VSI control could affect the stability of power grids. In addition, the stability of grid-connected VSI can be affected by a weak grid. A grid that has low Short Circuit Ratio (SCR) is said to be weak. In other words, a grid is said to be weak if it has impedance that is high and low inertia constant. In [12], the controllers of wind turbines were enhanced due to variations in weak grids. Muljadi et al. [13] investigated the control structure of a PMSG wind turbine in a weak grid, where the control of the DC-link was done by the Machine Side Converter (MSC).

This chapter targets the improved performance of PMSG-based wind generators in relation to grid codes, considering a hybrid scheme of a Series Dynamic Braking Resistor (SDBR) and line parameters of a weak grid. The effects of the placement of the SDBR with respect to the architecture of the PMSG wind generator machine and network side converters were investigated. The voltage of the grid terminal was used as the switching signal during transient condition, to control the SDBR. Furthermore, this chapter proposes a solution to some of the challenges in weak grids, by employing the proposed best-sized SDBR for the PMSG grid side, to improve its performance during transient state. The mathematical dynamics of the SDBR connection at the MSC of the PMSG wind turbine were analyzed. The mathematical model of the grid converter side of the PMSG wind turbine was also analyzed, considering the

DOI: 10.1201/9781003350910-6

impact of the SDBR insertion during transient state. The best-sized SDBR was used in the evaluation of the PMSG wind turbine at different network strengths, considering scenarios that characterized the grid, based on reactive power and ohmic values (X/R) ratio of the grid impedance (Z). The robustness of the PMSG wind turbine controllers was tested using both severe fault conditions of three-phase-to-ground and asymmetric faults.

6.2 MODELING OF THE PMSG WIND TURBINE

The modeling of the PMSG wind turbine can be referred to in Chapter 1 of this book.

6.3 THE MODEL SYSTEM AND CONTROL STRATEGY OF THE PMSG

The model system of study for this work is shown in Figure 6.1, where the full back-to-back power converters, transmission line and a step-up transformer are linking the PMSG wind turbine to an infinite bus, with a terminal voltage of 1.0 pu. The parameters of the model system are given in Table 6.1 [14]. The base system is 5.0

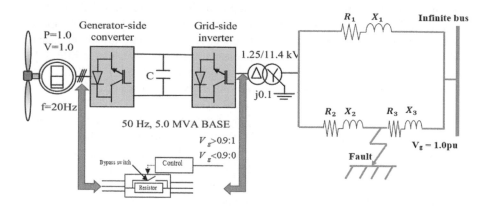

FIGURE 6.1   Model system of study with series dynamic braking resistor. (a) The SDBR at stator side of PMSG wind turbine (b) Topology of PMSG grid side voltage source converter.

TABLE 6.1
Parameters of the Model System

Rated power	5.0 MW	Rated voltage	1.0 kV
Rated voltage	1.0 kV	Field flux	1.4 pu
Frequency	20.0 Hz	Blade radius	40.0 m
Number of poles	150.0	Rated wind speed	12.0 m/s
Machine inertia	3.0	R_1	0.87120 Ω
Stator resistance	0.01 pu	R_2	0.04356 Ω
d-axis reactance	1.0 pu	R_3	0.82764 Ω
q-axis reactance	0.7 pu	X_1	5.2157 Ω
X_2	0.2608 Ω	X_3	4.9549 Ω

MVA, with a short circuit MVA of 16.67. A severe case three-phase-to-ground fault is considered to occur on the double circuit, as shown in the model system for the system evaluation performance. An SDBR is connected to either the MSC or the GSC of the PMSG wind turbine, as shown in the model system. Connecting the SDBR to the wind generator would improve its transient performance. Investigation on the optimal placement of the SDBR position in either the MSC or the GSC was carried out in subsequent sections of this chapter. The control scheme for the PMSG is already described in Chapters 4 and 5 of this book.

6.4 MATHEMATICAL DYNAMICS OF CONNECTING THE SDBR IN THE PMSG WIND TURBINE

The SDBR dynamics in the PMSG-based wind turbine are given in Figure 6.2(a). The distinctive merit of the SDBR topology lies in its operation principle, which is based on current and not voltage magnitude [15, 16]. During normal operation of the wind generator, the resistor is bypassed, because the SDBR switch is on. However, during

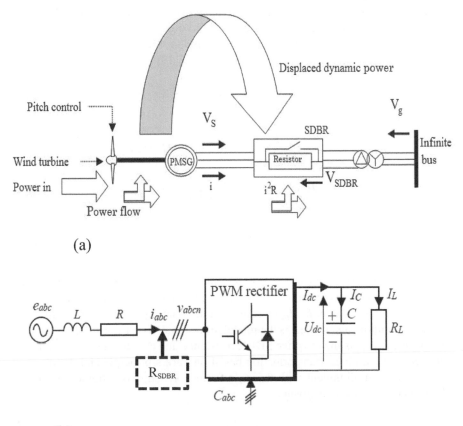

FIGURE 6.2 Dynamics of PMSG and SDBR.

fault conditions, the switch is off because of the series connection of the resistor. The switching signal used in switching the SDBR in this work is the terminal grid voltage, as given in Figure 6.1. Based on this configuration, the high rotor inrush current will be reduced and during transient conditions, the excessive wind turbine active power expected will be dissipated by the SDBR system [17, 18]. Consequently, there will be an effective balance between the MSC and GSC power converters. Besides, the proposed SDBR incorporation in the PMSG would limit the stator current and reduce the DC-link current capacitor charging. These activities would help in the mechanical system augmentation during transient states of the wind generator.

The main merits of the voltage source converter topology are high-quality DC-link voltage, bidirectional power flow capability, unity power factor and low AC-line current distortions. Since the PMSG-based wind turbine is decoupled fully from the network due to the back-to-back power converter, the SDBR can be related to the grid-tied inverter dynamics, as shown in Figure 6.2(b). From Figure 6.2(b), the PMSG-based wind turbine GSC is connected to the R and L parameters of the grid, with AC currents $i = a,b,c$. C_{abc} represents the three states of switching for the insulated gate bipolar transistors (IGBTs). The C_{abc} converter functions can be substituted by β_{abc} signals of modulation during high PWM switching frequency. Based on Park's transformation, the PMSG grid-connected voltage source converter could be modeled in a rotating frame. The mathematical model for the balanced three-phase voltage source converter in Figure 6.2(b) is given as [19]:

$$e_d = -\omega L i_q + L\frac{di_d}{dt} + (R + R_{SDBR})i_d + 0.5U_{dc}\beta_d \qquad (6.1)$$

$$e_q = -\omega L i_d + L\frac{di_q}{dt} + (R + R_{SDBR})i_q + 0.5U_{dc}\beta_q \qquad (6.2)$$

$$C\frac{dU_{dc}}{dt} = 0.75(i_d\beta_d + i_q\beta_q) - \frac{U_{dc}}{R_L} \qquad (6.3)$$

$$r = \sqrt{\beta_d^2 + \beta_q^2} \qquad (6.4)$$

From Equations (6.1) to (6.4), i_d, i_q represent the dq current input of the rectifier's axes, e_d, e_q are known as the dq voltage of the grid voltage axes components, ω is the angular frequency voltage, β_d, β_q represent the modulating signal of the rectifier's d and q axes components, while r is the modulation signal vector norm, U_{dc},is the DC-link voltage, R_{SDBR}, is the effective SDBR resistance and ω is the angular frequency. Considering three-phase transformation based on Park's principle and the phase-A grid voltage is in alignment with the dq reference synchronous frame, the source voltage dq components are given as:

$$e_d = E_m \qquad (6.5)$$

$$e_q = 0 \qquad (6.6)$$

From Equation (6.5), E_m is the phase grid voltage amplitude, e_d and e_q are the d and q source voltages. Consequently, the fed active (P_s) and reactive (Q_s) rectifier's powers are computed by

$$P_s = \frac{3}{2} E_m i_d \tag{6.7}$$

$$Q_s = -\frac{3}{2} E_m i_q \tag{6.8}$$

To achieve power factor in unity mode of operation, i_{q_ref} can be set to 0. Therefore, for the current regulation to be idea, $i_q = i_{qref} = 0$. With $i_q = 0$ and $e_q = 0$, the mathematical model of the voltage source converter under unity power factor can be expressed with the following set of equations:

$$E_m = L \frac{di_d}{dt} + \left(R + R_{SDBR} \right) i_d + 0.5 U_{dc} \beta_d \tag{6.9}$$

$$\beta_q = -\frac{2\omega L}{U_{dc}} i_q \tag{6.10}$$

$$C \frac{dU_{dc}}{dt} = \frac{3}{4} i_d \beta_d - \frac{U_{dc}}{R_L} \tag{6.11}$$

Equation (6.10) implies in order to ensure operation of unity power factor of the voltage source converter, the component of β_q should proportionally vary with i_q current. From Equations (6.9) and (6.11), the capacity charge is manipulated by β_d, via the i_d current of the input.

The insertion of the SDBR resistance during transient state in the PMSG converter would affect the maximal power flow and the DC output voltage. During normal condition, the derivative operator relating all terms in Equations (6.9)–(6.11), would be zero. Thus, the new set of equations would be:

$$E_m = \left(R + R_{SDBR} \right) i_d + 0.5 U_{dc} \beta_d \tag{6.12}$$

$$\beta_q = -\frac{2\omega L}{U_{dc}} i_q \tag{6.13}$$

$$i_d = \frac{4 U_{dc}}{3 \beta_d R_L} \tag{6.14}$$

Putting Equation (6.14) into (6.12), for a given load of R_L and voltage U_{dc}, the expression of the signal command β_d is

$$6 E_m R_L \beta_d - 8 \left(R + R_{SDBR} \right) U_{dc} - 3 R_L \beta_d^2 U_{dc} = 0, \text{ for } \beta_d \neq 0 \tag{6.15}$$

There are two derived solutions from Equation (6.15):

$$\beta_{d1} = \frac{E_m}{U_{dc}} - \sqrt{\left(\frac{E_m}{U_{dc}}\right)^2 - \frac{8(R + R_{SDBR})}{3R_L}} \tag{6.16}$$

$$\beta_{d2} = \frac{E_m}{U_{dc}} + \sqrt{\left(\frac{E_m}{U_{dc}}\right)^2 - \frac{8(R + R_{SDBR})}{3R_L}} \tag{6.17}$$

Since β_d component in Equation (6.16) has very low values, this solution is not admissible. Therefore, the solution of Equation (6.17) is the acceptable solution, making $\beta_d = \beta_{d2}$. And β_d would exist based on the following condition been satisfied:

$$\left(\frac{E_m}{U_{dc}}\right)^2 - \frac{8(R + R_{SDBR})}{3R_L} \geq 0 \tag{6.18}$$

The above condition is with respect to power consumption by the load. The operation of the converter is possible if

$$P_{dc} \leq P_{dc\_max} \tag{6.19}$$

where P_{dc_max} is the maximal available power at the DC-connected side power of the PMSG converter. Considering the power conservation principle, P_{dc_max} can be expressed as

$$P_{dc} = \frac{3}{2}E_m i_d - \frac{3}{2}(R + R_{SDBR})i_d^2 \tag{6.20}$$

By solving $dP_{dc} / di_d = 0$, the maximum power transfer to the DC side is

$$\frac{dP_{dc}}{di_d} = \frac{3}{2}E_m - 3Ri_d = 0 \rightarrow i_d = i_{d\_max} = \frac{E_m}{2R} \tag{6.21}$$

Putting Equation (6.21) into (6.20) yields the maximum DC side power as,

$$P_{dc\_max} = \frac{3E_m^2}{8(R + R_{SDBR})} \tag{6.22}$$

The PMSG voltage source converter operation is possible when

$$P_{dc} \leq P_{dc\_max} \rightarrow \left(\frac{E_m}{U_{dc}}\right)^2 - \frac{8(R + R_{SDBR})}{3R_L} \geq 0 \tag{6.23}$$

Equation (6.23) reflects the condition of Equation (6.18). Finally, the grid input maximal power P_{s_max} can be obtained by putting Equation (6.22) into (6.8) for Q_s. Thus:

$$P_{s\_max} = \frac{3E_m^2}{4\left(R + R_{SDBR}\right)} \tag{6.24}$$

Based on Equation (6.24), the maximum power transfer of the GSC of the PMSG during transient state would be reduced by the insertion of SDBR in the GSC. Therefore, the total current reduces its value, thus, mitigating the oscillations that normally occur during transient conditions.

6.5 DYNAMICS OF WEAK GRIDS AND VOLTAGE SOURCE INVERTERS

The control of frequency converters could be very challenging when considering the wind generators' penetration because of the grid parameters uncertainty. The stability of the grid would be affected as the number of inverters increases. In weak grids, voltage fluctuations and inverter instability are bound to occur [20, 21]. With the connection of an active load like the PMSG with rated power S_{PMSG} into the network, the SCR could be expressed in terms of Short Circuit Capacity (SCC) as:

$$\text{Short circuit ratio} = \frac{SCC}{S_{PMSG}} = \frac{U_G^2}{Z_{weak} \cdot S_{PMSG}} \tag{6.25}$$

where the SCC is the amount of flowing power at the point of a short circuit, U_G is the grid voltage and Z_{weak} is the impedance of the weak grid. As the standard, with SCR lower than factor of 10, the grid is said to be weak [12]. The grid could be characterized also by considering reactive power and ohmic values of its impedance (Z). This is known as the X/R ratio or (xrr). Equation (6.26) shows the relationship between (Z), xrr, the inductive (X) and ohmic amount (R) for a typical grid system. Usually, in weak grids, xrr is low, about 1/2, therefore, these types of grids have ohmic features.

$$X = \frac{Z}{\sqrt{1+\left(\dfrac{1}{xrr}\right)^2}} \tag{6.26a}$$

$$R = \frac{Z}{\sqrt{1+\left(xrr\right)^2}} \tag{6.26b}$$

The equivalent circuit of a typical weak grid model is given in Figure 6.3 [22]. In this circuit, the voltage of the grid U_G is said to be constant and the PMSG is tied at

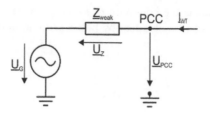

FIGURE. 6.3   Circuit of a weak grid model.

the Point of Common Coupling (PCC) to the weak grid, injecting a current of I_{WT} with a voltage of U_{PCC}. This U_{PCC} is referred to as the a reference voltage.

The complex power at the PCC with the PMSG wind turbine only connected is given by [21]:

$$S_{PCC} = U_{PCC}.I_{WT}^* = U_{PCC}.\left(\frac{U_{PCC}-U_G}{Z_{weak}}\right)^* = P + jQ \qquad (6.27)$$

Expanding Equation (6.27) gives:

$$\left(U_{PCC}^2\right)^2 - U_{PCC}^2.\left(U_G^2 + 2RP + 2XQ\right) + \left(P^2 + Q^2\right)\left(R^2 + X^2\right) = 0 \qquad (6.28)$$

From Equation (6.28), there are two U_{PCC} solutions. The first solution describes the stable or possible solution, while the second describes the impossible or unstable operation of the grid. In the first solution, U_{PCC} is the voltage with a value of 1.0 pu, with a well-defined frequency and amplitude. However, this is not the case for the second solution. When the PMSG wind generator injects its full power $S_{PMSG} = 1.0$ pu into the weak grid, U_{PCC} voltage would change for different P and Q combinations. Because the PMSG wind generator will always inject active power, P will be positive always, while the inductive or capacitive elements give the reactive power. In light of the above, for a very weak grid, a stable operation cannot be achieved, without external compensation or oversized GSC. Also, if there is a possibility of power injection into the grid, there will always be a PQ combination, in order to achieve fixed voltage at the PCC.

This chapter uses SDBR in the PMSG wind turbine GSC to overcome this issue. Based on Equations (6.1)–(6.24), the insertion of the SDBR in the GSC of the PMSG would enhance its responses as would be demonstrated in Section 6.6.2 in the evaluation of the system performance.

6.6   RESULTS AND DISCUSSIONS

6.6.1   THE PLACEMENT AND EFFECTIVE SIZING OF THE SDBR IN PMSG-BASED WIND TURBINE

The system performance was evaluated using PSCAD/EMTDC [23] environment, with the considered model system of Figure 6.1. The fault type is a severe three-phase of 100 ms happening at 10.1 s, with the circuit breaker's operation sequence opening and reclosing at 10.2 s and 11 s, respectively, on the faulted line. The first part of the evaluation of the system performance was carried out considering the position of the SDBR at the MSC and the GSC, respectively. A scenario where SDBR was not placed at the PMSG wind turbine was also investigated. The wind speed used for the PMSG wind generator during operation was its rated value. Figures 6.4–6.8 show the performances of the various variables of the wind turbine. From the responses of these figures, connecting the SDBR to the MSC of the wind generator has no major effect on the PMSG during transient state. This is because the PMSG-based wind turbine is decoupled fully from the network using a back-to-back power converter. The undershoot, overshoot and settling time of the terminal voltage of the wind turbine variables with the SDBR connected to the MSC was more in Figure 6.4, compared to other scenarios. However, placing the SDBR on GSC has a great influence on the wind generator variables in fault conditions. This is because the expected wind generator stator circuitry of the PMSG high voltage is divided by the SDBR since its connection is series topology. In Figures 6.4 and 6.5, the undershoot, overshoot and time of settling for the terminal voltage and DC-link voltage were much improved considering the effectively sized SDBR. Consequently, there is no loss in control of the power converters, since no induced overvoltage would be experienced. More so,

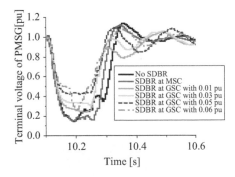

FIGURE 6.4   Terminal voltage of PMSG wind turbine.

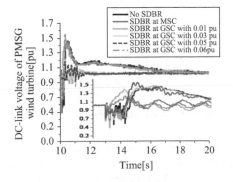

FIGURE. 6.5   DC-link voltage of PMSG wind turbine.

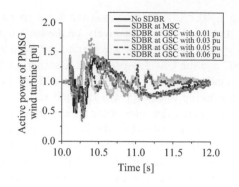

FIGURE 6.6   Active power of PMSG wind turbine.

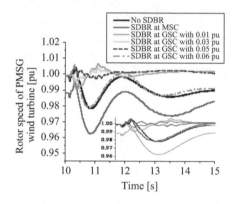

FIGURE 6.7   Rotor speed of PMSG wind turbine.

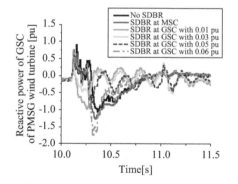

FIGURE 6.8   Reactive power of PMSG wind turbine.

during operation, this solution of FRT has the capability to mitigate high current flow significantly, thus, avoiding dangerous overvoltage and high charging current that is normally experienced at the DC-based capacitor in Figure 6.5 power converters that are fragile in nature.

The salient aspect of the chapter is the effective sizing investigation for the SDBR in the PMSG wind turbine GSC. From Figures 6.4 to 6.8, a too-small or large value of SDBR would result in delay in the recovery of the variables of the wind turbine after transient conditions. An effective size of SDBR 0.05pu gave improved responses of the PMSG variables. Figures 6.4–6.6 show the responses of the wind generator terminal voltage, for no scenario, with SDBR at the MSC and with SDBR of various values at the GSC of the wind generator. With no SDBR, poor performance of the PMSG wind turbine was experienced especially for the terminal voltage, with high undershoot, overshoot and more settling time.

Connecting the SDBR at the MSC converter, apparently, did not enhance the variables of the PMSG as earlier stated. However, connecting the various values of SDBR has a great effect on the terminal voltage, DC-link voltage and the PMSG active power, respectively. The DC-link voltage was improved, during transient condition in Figure 6.8, thereby enhancing the active power in Figure 6.6 and the rotor speed of the PMSG wind turbine in Figure 6.7, with lower values of undershoot, overshoot and time of settling of the variables considering the effective-sized SDBR. This is because the SDBR control technique in the PMSG can increase its mechanical output, thus, the transient state condition excursion speed would be limited. Consequently, the speed performance would have reduced oscillations, as shown in Figure 6.7, with lower undershoot, overshoot and faster settling time. Also, because the SDBR has the ability to boost the reactive power dissipation of the GSC of the wind generator in Figure 6.8, the terminal voltage of the grid would definitely be enhanced as a result of the direct relation between reactive power and voltage, as presented in Figure 6.4.

A small effective size of SDBR of value 0.05 pu enhanced the performance of the PMSG variables, compared to a smaller size of 0.01 pu or a larger size of 0.06 pu. The effective SDBR size is further used for the investigation of the effect of weak grid line parameters on the response of the PMSG variables considering SDBR at the GSC in Section 6.6.2.

6.6.2 IMPROVING THE PERFORMANCE OF PMSG WIND TURBINE IN WEAK GRIDS CONSIDERING THE EFFECTIVE SIZE OF SDBR

The same model system and fault condition were considered in this section, in order to effectively judge the performance of the system. Five scenarios with different weak grid parameters, as shown in Table 6.2, were carried out. The effective SDBR size of 0.05 pu was used in this analysis. Figures 6.9–6.13 show the response of the PMSG wind turbine variables. The values of the line parameters are given in Table 6.1, while their X/R ratios are given in Table 6.2. The line parameters were varied for five scenarios, in order to know their effects on the PMSG system. Furthermore, the effective-sized SDBR was used to enhance the PMSG wind turbine performance. From Table 6.2, scenarios 1 and 2 were designed for very weak grids with and without considering the effective SDBR in the PMSG-based wind turbine. Scenario 2 is the weakest grid but with SDBR and scenario 1 is the weakest grid but without SDBR. From Figure 6.9, scenario 2 gives better results than scenario 1 in the weakest grid scenario, thus, it is justifiable to use SDBR as a recommendation for weak grids.

TABLE 6.2

Line Parameters of the Network

Scenarios	Effective SDBR	$X/R(\Omega)$
1	0	4.0
2	0.05 pu	4.0
3	0.05 pu	6.0
4	0.05 pu	8.0
5	0.05 pu	10.0

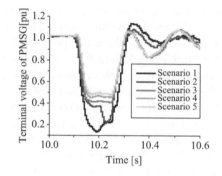

FIGURE 6.9   Terminal voltage of PMSG with network parameters.

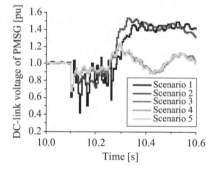

FIGURE 6.10   DC-link voltage of PMSG with network parameters.

The undershoot, overshoot and faster time of settling for scenario 2 is better than scenario 1 for the terminal voltage variable. A further increase of the X/R ratio for scenario 5 makes the grid strong and with the addition of SDBR, better performance is obtained with lower overshoot, undershoot and faster settling time. The voltage of the DC bus along with the response of the reactive power for the PMSG variables was improved in Figures 6.10 and 6.11, respectively, for scenario 5 for the strongest grid with SDBR, compared to the weakest grid with or without SDBR. Though few oscillations were observed for these variables in the weakest grid with SDBR, however, their overshoot and undershoot were slightly more than the scenario without SDBR.

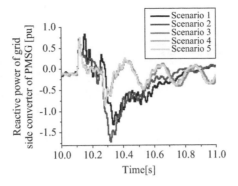

FIGURE 6.11   Reactive power of GSC of PMSG with network parameters

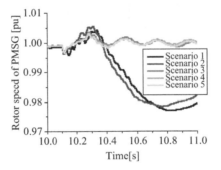

FIGURE 6.12   Rotor speed of PMSG with network parameters.

This is compensated for in the response of the terminal voltage because more reactive power was dissipated. Figure 6.12 reflects that improvement could be achieved via the use of the SDBR for the rotor speed in the weakest grid and much better performance in scenario 5 during transient state.

6.6.3   ANALYSIS OF THE PROPOSED SCHEME CONSIDERING ASYMMETRICAL FAULTS

6.6.3.1   Double-Line-to-Grid Fault (2 LG Fault)

The chapter also carried out the investigation of double-line-to-ground fault using the same model system employed for the three-line-to-ground fault earlier discussed, in order to conclude that the proposed SDBR PMSG scheme could perform better during transient state. Figure 6.13 (a–f) shows the performance of the PMSG variables during double-line-to-ground fault. In Figure 6.13 (a–c), the responses of the terminal voltage, DC-link voltage and rotor speed of the PMSG were better with the SDBR having an effective value of 0.5 pu connected to the GSC of the wind generator, instead of the MSC. Furthermore, considering the strength of the grid, the responses in Figure 6.13 (d–f) show that the proposed scheme can improve the performance of the PMSG wind turbine in weak grids, when a double-line-to-grid fault is applied to the model system.

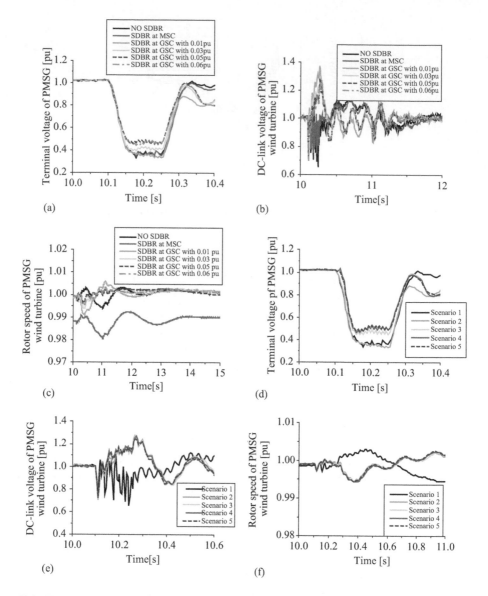

FIGURE 6.13 Response of the PMSG wind turbine for 2 LG asymmetrical fault. (a) Terminal voltage of PMSG 2 LG. (b) DC-link voltage of PMSG 2 LG. (c) Rotor speed of PMSG 2 LG. (d) Terminal voltage of PMSG 2 LG for all scenarios. (e) DC-link voltage of PMSG 2 LG for all scenarios. (f) Rotor speed of PMSG 2 LG for all scenarios.

6.6.3.2 Line-to-Line Grid Fault (LL Fault)

Similarly, the line-to-line fault was investigated, using the same model system for the three-line-to-ground fault, to check the performance of the proposed SDBR PMSG scheme during transient state. Figure 6.14 (a–f) shows the performance of the PMSG variables during line-to-line fault. In Figure 6.14 (a–c), the responses of the terminal voltage, DC-link voltage and rotor speed of the PMSG were better with the SDBR

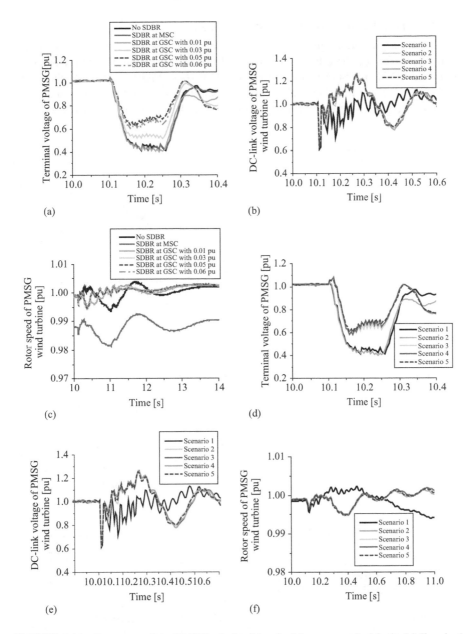

FIGURE 6.14 Response of the PMSG wind turbine for LL asymmetrical fault. (a) Terminal voltage of PMSG LL. (b) DC-link voltage of PMSG LL. (c) Rotor speed of PMSG LL. (d) Terminal voltage of PMSG LL for all scenarios. (e) DC-link voltage of PMSG LL for all scenarios. (f) Rotor speed of PMSG 2 LL for all scenarios.

having an effective value of 0.5 pu connected to the GSC of the wind generator, instead of the MSC. Furthermore, considering the strength of the grid, the responses in Figure 6.14 (d–f) show that the proposed scheme can improve the performance of the PMSG wind turbine in weak grids when subjected to a line-to-line fault.

6.6.3.3   Line-to-Ground Grid Fault (1 LG Fault)

Finally, a single-line-to-ground fault was considered in the same model system earlier used for the three-line-to-ground fault, to help ascertain the behavior of the proposed SDBR PMSG scheme during transient state. Figure 6.15 (a–f) shows the performance of the PMSG variables during a single-to- ground fault. In Figure 6.15 (a–c), the responses of the terminal voltage, DC-link voltage and rotor speed of the

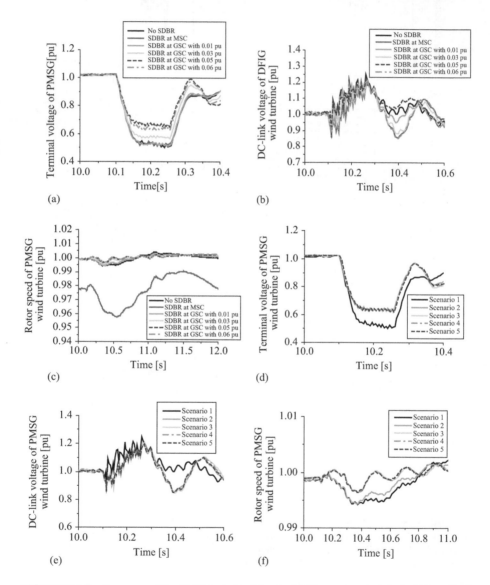

FIGURE 6.15   Response of the PMSG wind turbine for 1 LG asymmetrical fault. (a) Terminal voltage of PMSG 1LG. (b) DC-link voltage of PMSG 1LG. (c) Rotor speed of PMSG 1LG. (d) Terminal voltage of PMSG LL for all scenarios. (e) DC-link voltage of PMSG LL for all scenarios (f) Rotor speed of PMSG 2 LG for all scenarios.

PMSG were better with the SDBR having an effective value of 0.5 pu connected to the GSC of the wind generator, instead of the MSC. Furthermore, considering the strength of the grid, the responses in Figure 6.15 (d–f) show that the proposed scheme can improve the performance of the PMSG wind turbine in weak grids when subjected to a single-line-to-ground fault.

6.6.3.4 Comparison of the Various Asymmetric Fault Conditions

A comparative analysis was carried out using the proposed scheme that gave better performance (scenario 5) in Section 6.6.2 with the model system in Figure 6.1, considering double-line-to-ground, line-to-line and single-line-to-ground asymmetric faults, respectively. As shown in Figure 6.16, the terminal voltage of the PMSG experienced more voltage deep for 2LG, compared to the other asymmetric fault scenarios. Figure 6.17 shows that the response of the rotor speed of the wind generator was more affected considering the oscillations and settling time for 2LG, compared to the LL and 1 LG asymmetric fault scenarios, respectively. These results demonstrate the effectiveness and robustness of the controllers employed for the improved performance of the proposed PMSG wind generator scheme considering any fault scenario.

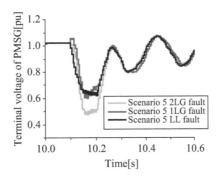

FIGURE 6.16   Terminal voltage of PMSG for asymmetrical faults.

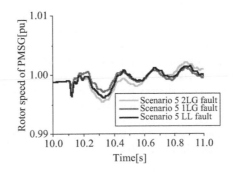

FIGURE 6.17   DC-link voltage of PMSG for asymmetrical fault.

6.7 CHAPTER CONCLUSION

As the penetration of a huge amount of wind energy in modern power grids is on the rise, failure to provide robust fault ride through solutions would lead to the continuous shutdown of power systems made of multi-machine and wind farms. One of the targets of this chapter was to enhance the performance of the PMSG wind turbine considering an SDBR, fault current limiter. The SDBR position was investigated by connecting it at the machine and grid side converters of the wind generator. From the study in this chapter, it was found out that placing SDBR at the PMSG grid side would enhance its performance, compared to connecting it at the machine side. A further investigation of the size of the SDBR was done in this chapter. When a small- or large-sized SDBR was used in the network side of the PMSG, the variables of the wind generator were delayed in recovery and settling back to their steady states during transient conditions, however, an effectively sized SDBR value gave a better performance. More so, the performance of the wind turbine using the effectively sized SDBR was investigated, considering different scenarios of the network line parameters. It was observed that since weak grids are characterized by their X/R ratio, the use of the SDBR scheme for the PMSG wind turbine would improve its performance by the selection of the appropriate resistive and reactance parameters of the grid. Thus, it is justifiable to use SDBR as a recommendation for weak grids transient performance enhancement.

REFERENCES

[1] M. Youjie, T. Long, Z. Xuesong, L. Wei, and S. Xueqi, "Analysis and control of wind power grid integration based on a permanent magnet synchronous generator using a fuzzy logic system with linear extended state observer," *Energies*, vol. 12, pp. 2862, 2019, doi: 10.3390/en12152862.

[2] T. Ramji, and B. N. Ramesh, "Fuzzy logic based MPPT for permanent magnet synchronous generator in wind energy conversion system," *IFAC-conference*, vol. 49, no. 1, pp. 462–467, 2016.

[3] X. Han, and Y. Ma, "Finite-time extended dissipative control for fuzzy systems with nonlinear perturbations via sampled-data and quantized controller," *ISA Transactions*, vol 89, pp. 31–44, 2019.

[4] R. Mohammadikia, and M. Aliasghary, "A fractional order fuzzy PID for load frequency control of four-area interconnected power system using biogeography-based optimization," *International Transactions on Electrical Energy Systems*, vol. 29, pp. 1–17, 2019.

[5] P. J. Gaidhane, M. J. Nigam, A. Kumar, and P. M. Pradhan, "Design of interval type-2 fuzzy precompensated PID controller applied to two-DOF robotic manipulator with variable payload," *ISA Transactions*, vol. 89, pp. 169–185, 2019.

[6] A. S. Mahmoud, M. H. Hany, Z. A. Haitham, E. E. El-Kholy, and A. M. Sabry, "An adaptive fuzzy logic control strategy for performance enhancement of a grid-connected PMSG-Based wind turbine," *IEEE Transactions on Industrial Informatics*, vol. 15, no. 6, pp. 3163–3173, 2019.

[7] N. P. W. Strachan, and D. Jovcic, "Stability of a variable-speed permanent magnet wind generator with weak ac grids," *IEEE Transactions on Power Delivery*, vol. 25, no. 4, pp. 2779–2788, 2010.

[8] J. Hu, Y. Huang, D. Wang, H. Yuan, and X. Yuan, "Modeling of grid-connected DFIG-based wind turbines for dc-link voltage stability analysis," *IEEE Transactions on Sustainable Energy*, vol. 6, no. 4, pp. 1325–1336, 2015.

[9] Y. Zhou, D. D. Nguyen, P. C. Kjr, and S. Saylors, "Connecting wind power plant with weak grid - challenges and solutions," *2013 IEEE Power Energy Society General Meeting*, pp. 1–7, 2013.

[10] J. M. Carrasco, L. G. Franquelo, J. T. Bialasiewicz, E. Galvan, R. C. P. Guisado, M. A. M. Prats, J. I. Leon, and N. Moreno-Alfonso, "Power-electronic systems for the grid integration of renewable energy sources: A survey," *IEEE Transactions on Industrial Electronics*, vol. 53, no. 4, pp. 1002–1016, 2006.

[11] F. Blaabjerg, Z. Chen, and S. B. Kjaer, "Power electronics as efficient interface in dispersed power generation systems," *IEEE Transactions on Power Electronics*, vol. 19, no. 5, pp. 1184–1194, 2004.

[12] X. Yuan, F. Wang, D. Boroyevich, Y. Li, and R. Burgos, "DC-link voltage control of a full power converter for wind generator operating in weak-grid systems," *IEEE Transactions on Power Electronics*, vol. 24, no. 9, pp. 2178–2192, 2009.

[13] E. Muljadi, C. P. Butterfield, B. Parsons, and A. Ellis, "Effect of variable speed wind turbine generator on stability of a weak grid," *IEEE Transactions on Energy Conversion*, vol. 22, no. 1, pp. 29–36, 2007.

[14] H. M. Hasanien, and S. M. Muyeen, "Design optimization of controller parameters used in variable speed wind energy conversion system by genetic algorithms," *IEEE Transactions on Sustainable Energy*, vol. 3, no. 2, pp. 200–208, 2012.

[15] K. E. Okedu, S. M. Muyeen, R. Takahashi, and J. Tamura, "Wind farms fault ride through using DFIG with new protection scheme," *IEEE Transactions on Sustainable Energy*, vol. 3 no. 2, pp. 242–254, 2012.

[16] K. E. Okedu, "Determination of the most effective switching signal and position of braking resistor in DFIG wind turbine under transient conditions," *Electrical Engineering*, vol. 102, no. 11, pp. 471–480, 2020.

[17] K. E. Okedu, S. M. Muyeen, R. Takahashi, and J. Tamura, "Wind farm stabilization by using DFIG with current controlled voltage source converters taking grid codes into consideration," *IEEJ Transactions on Power and Energy*, vol. 132, no. 3. pp. 251–259, 2012.

[18] K. E. Okedu, S. M. Muyeen, R. Takahashi, and J. Tamura, "Improvement of fault ride through capability of wind farm using DFIG considering SDBR," *14th European Conference of Power Electronics EPE*, Birmingham, UK, August 2011, pp. 1–10.

[19] M. Koussaila, O. C. Tahar, R. Adel, and S. Hamid, "Voltage source converter parameters design considering equivalent source resistance effect," *The 5th International Conference on Electrical Engineering–Boumerdes (ICEE-B)*, Boumerdes, Algeria, 29–31 October 2017.

[20] D. Yang, X. Ruan, and H. Wu, "Impedance shaping of the grid-connected inverter with LCL filter to improve its adaptability to the weak grid condition," *IEEE Transactions on Power Electronics*, vol. 29, no. 11, pp. 5795–5805, 2014.

[21] C. V. Thierry, *Voltage Stability of Electric Power Systems*, New York, US: Springer, 1998.

[22] S. Grunau, and W. F. Fuchs. "Effect of wind-energy power injection into weak grids," *Institute for Power Electronics and Electrical Drives, Christian-Albrechts-University of Kiel D-24143* Kiel, Germany, pp. 1–7, 2012.

[23] PSCAD/EMTDC Manual, Manitoba HVDC lab, 2016.

7 DFIG and PMSG in Weak and Strong Grids

7.1 CHAPTER INTRODUCTION

Power grids could be classified based on their Short Circuit Ratio (SCR). A power grid that has low SCR is said to be weak, while a power grid with high SCR is said to be strong. In other words, a power grid is said to be weak if it has an impedance that is high and inertia low constant, while a power grid is said to be strong if it has an impedance that is low and inertia high constant. In the literature, several works on weak grids and the challenges they pose on network stability considering wind energy penetration has been reported [1–3]. There is a great influence of interfacing a Voltage Source Converter (VSC) with Pulse Width Modulated (PWM) on the power grid stability. Garcia-Garcia et al. [4] showed that the controllers of wind turbines could be improved in weak grids during voltage variations.

In this chapter, a comparative analysis regarding stability issues faced by DFIG and PMSG wind turbines is investigated in various grid strengths. The modeling and wind turbine characteristics of both wind generators were analyzed along with their control strategies. Both wind turbines were subjected to weak, normal and strong grids with a severe bolted three-line-to-ground fault, without any protection or enhancement scheme, in order to test the robustness of the controllers employed. The mathematical dynamics of the implementation of the series dynamics braking resistor (SDBR) as a limiter of the fault current at the stator of both wind turbines for effective comparison study was also presented. The effective SDBR size was used for both wind turbine technologies, considering the same switching signal of the grid voltage during transient state. Furthermore, an overvoltage protection system was considered for both wind turbines based on the connection of a DC chopper circuitry at the power converter of the wind turbines. A combination of the SDBR and overvoltage protection scheme of the DC chopper was employed for both wind turbines operating at weak grids, which is characterized based on reactive power and ohmic values (X/R) ratio of the grid impedance (Z). The evaluation of both wind turbines was done for weak, normal and strong grids, considering the same machine ratings of the wind turbines. Because of the critical situations of the wind turbines during faulty conditions in the weak grids, an analysis was done considering the use of effective Series Dynamic Braking Resistor (SDBR) for both wind turbines. The grid voltage variable was employed as the signal for switching the SDBR in both wind turbines during transient state. Also, an overvoltage protection system was considered for both wind turbines using the DC chopper in the DC-link excitation circuitry of both wind turbines. Furthermore, a combination of the SDBR and DC chopper was employed in both wind turbines at weak grid conditions in order to improve the performance of the variable speed wind turbines. There is a limited number of studies

DOI: 10.1201/9781003350910-7

in the literature that considered the scenarios of weak and strong grids for both wind turbines, in conjunction with the SDBR control and overvoltage protection topologies. Most studies in the literature considered these scenarios on a separate basis for the fault ride through enhancement of both wind generators.

7.2 MODELING AND CONTROL

7.2.1 WIND TURBINE CHARACTERISTICS

The DFIG torque and mechanical extracted power are given as follows in Equations (7.1) and (7.2), respectively [4, 5].

$$T_{\mathrm{m}} = \frac{\pi \rho R^3}{2} V_{\mathrm{w}}^2 C_t(\lambda) \left[Nm \right] \tag{7.1}$$

$$P_{\mathrm{m}} = \frac{\pi \rho R^2}{2} V_{\mathrm{w}}^3 C_P(\lambda) \left[W \right] \tag{7.2}$$

where ρ: air density, R: radius of the turbine, V_{w}: wind speed, $C_p(\lambda, \beta)$, power coefficient is given by

$$C_p(\lambda, \beta) = 0.5 \left(\Gamma - 0.02 \beta^2 - 5.6 \right) e^{-0.17\Gamma} \tag{7.3}$$

The relationship between C_t and C_P is

$$C_t(\lambda) = \frac{C_p(\lambda)}{\lambda} \tag{7.4}$$

$$\lambda = \frac{\omega_r R}{V_{\mathrm{w}}} \tag{7.5}$$

$\Gamma = \dfrac{R\,(3600)}{\lambda\,(1609)}$ from Equation (7.3) and the ratio of the speed tip is λ.

The wind turbine characteristics are shown in Figure 7.1. The range of the rotor speed for DFIG is between 0.7 p.u and 1.3 p.u and maximum power obtains a turbine speed of 0.97 pu. On the other hand, maximum power for PMSG is obtained when the turbine speed is 1.0 pu.

7.2.2 DFIG MODEL AND CONTROL

Details of the DFIG model and control can be obtained from Chapters 1, 2, 3 and 5 of this book [6–9]. The Rotor Side Converter (RSC) control of the DFIG in Figure 7.2 is regulated by the axes q and d currents i_{qr} and i_{dr}. This is done by (P_s, Q_s), of the stator. The MPPT obtains the P_s. In Figure 7.3, the Grid Side Converter (GSC) of the wind generator employs reference frame ac grid to control the DC-link voltage and flow of reactive power exchange at the Point of Common Coupling (PCC) based on power flow direction of the rotor circuitry. The dq-to-abc and abc-to-dq

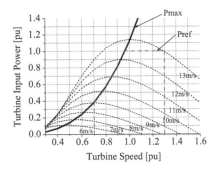

FIGURE 7.1 The wind turbine characteristics.

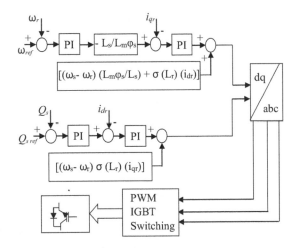

FIGURE 7.2 The rotor side converter control circuit of the DFIG.

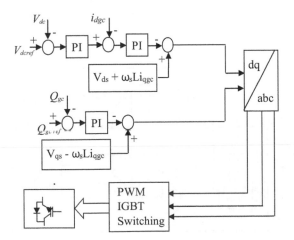

FIGURE 7.3 The grid side converter control circuit of the DFIG.

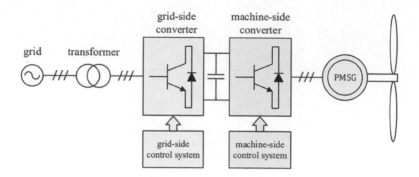

FIGURE 7.4   PMSG-based wind turbine.

transformation regarding the ac voltage synchronism is achieved with the help of an in-built phase lock loop model.

7.2.3   PMSG Model and Control

The PMSG reference power P_{ref} is limited to wind generator-rated power. The full back-to-back power converters of the PMSG turbine [10], tied to the grid, are shown in Figure 7.4. For Maximum Power Point Tracking (MPPT) realization [11, 12], the speed control is achieved by the Machine Side Converter (MSC). However, the GSC takes care of the effective voltage of the DC stabilization and regulation of the power factor and power quality.

Details of the PMSG model and control can be obtained in Chapters 1, 4, 5 and 6 of this book [13–16]. Figure 7.5 shows the controller of the MSC of the PMSG, which regulates both the output power of the active and reactive variables. The abc-to-dq-based variables transformation is achieved via angle position rotor (θ_r) considering the rotor speed of the wind generator. The I_{sd}, I_{sq} regulates both (P_s), (Q_s) of the wind generator, respectively. The MPPT technique is employed in the (P_{ref}). Usually (Q_s^*) is fixed at 0, to obtain effective operation of power factor 1. Three-phase

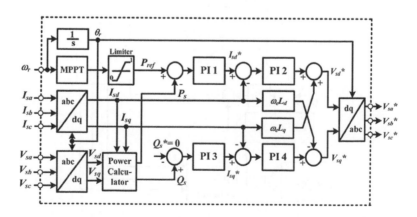

FIGURE 7.5   The rotor side converter control circuit of the PMSG.

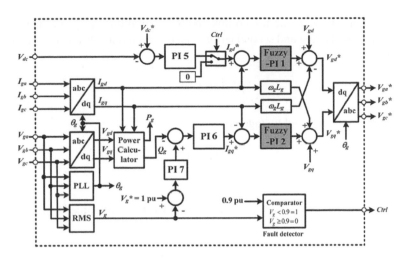

FIGURE 7.6 The grid side converter control circuit of the PMSG.

reference voltages (V_{sa}^*, V_{sb}^*, V_{sc}^*) are generated for the PWM switching, through the outputs of the current controller, based on the voltage references V_{sd}^* and V_{sq}^*. The GSC control in Figure 7.6 is regulated considering the d-q rotating reference frame, based on the voltage of the grid and the speed of rotation. The Park transformation is used in converting (I_{ga}, I_{gb}, I_{gc}) and (V_{ga}, V_{gb}, V_{gc}) into their rotating reference d-q frame. The extraction of the phase angle (θ_g) is done by considering the phase lock loop.

7.3 MATHEMATICAL DYNAMICS OF SDBR IN DFIG WIND TURBINE

The stator voltage of the DFIG wind turbine during transient conditions [17] is.

$$v_s^s = V_{s+}e^{j\omega_s t} + V_{s-}e^{-j\omega_s t} \tag{7.6}$$

V_{s+}, V_{s-} are components of the stator's voltage positive and negative sequences. At normal state, the stator flux is:

$$\psi_{ss}^s = \frac{V_{s+}e^{j\omega_s t}}{j\omega_s} + \frac{V_{s-}e^{-j\omega_s t}}{-j\omega_s} \tag{7.7}$$

With a sudden drop in grid voltage, there will be transient components in the stator flux to counteract transition in the state variable [18]. Therefore, the stator flux would have the natural flux (ψ_{sn}) expressed as:

$$\psi_s^s = \psi_{ss}^s + \psi_{sn}^s = \frac{V_{s+}e^{j\omega_s t}}{j\omega_s} + \frac{V_{s-}e^{-j\omega_s t}}{-j\omega_s} + \psi_{sn}^s e^{\frac{-t}{\tau_s}} \tag{7.8}$$

$\tau_s = \dfrac{L_s}{R_s}$ is the time constant of the stator flux.

During transient state, the forced flux and the natural flux (ψ_{sn}) in Equation (7.8) occur, which are the first and second terms. The rotor reference frame is related to the stator flux by

$$\psi_s^r = \psi_s^s e^{-j\omega_r t} \tag{7.9}$$

While the rotor-induced voltage is

$$\bar{v}_{ro} = \frac{L_m}{L_s} \frac{d\overline{\psi}_s}{dt} \tag{7.10}$$

The open circuit rotor voltage of the DFIG wind turbine is obtained from Equations (7.8) to (7.10) as:

$$\bar{v}_{ro} = \frac{L_m}{L_s} s V_+ e^{js\omega_s t} + \frac{L_m}{L_s}(s-2) V_- e^{j(2-s)\omega_s t} + \frac{L_m}{L_s}\left(j\omega_r + \frac{1}{\tau_s}\right)\psi_{sn} e^{-\frac{t}{\tau_s}} e^{j\omega_r t} \tag{7.11}$$

Neglecting the $\dfrac{1}{\tau_s}$ term in Equation (7.11) leads to:

$$\bar{v}_{ro} = \frac{L_m}{L_s} s V_+ e^{js\omega_s t} + \frac{L_m}{L_s}(s-2) V_- e^{j(2-s)\omega_s t} + \frac{L_m}{L_s} A(1-s) e^{-\frac{t}{\tau_s}} e^{j\omega_r t} \tag{7.12}$$

During grid fault, the DFIG stator flux is comprised of forced component and flux natural, with high rotor voltage transient and natural flux component. However, there is a decay of the natural flux with time constant $\tau_s = \dfrac{R_s}{L_s}$ of the stator circuit. Consequently, the stator resistance would increase due to the SDBR as below:

$$R_{s\,\text{effective}} = R_s + R_{sdbr} \tag{7.13}$$

The new time constant would now be $\tau_{s\,\text{effective}} = \dfrac{L_s}{R_{s\,\text{effective}}}$, making the total current lower, with reduced oscillations during transient state.

7.4 MATHEMATICAL DYNAMICS OF SDBR IN PMSG WIND TURBINE

Based on Park's transformation, the PMSG grid-connected VSC could be modeled in rotating frame. A recap of the mathematical model for the balanced three-phase VSC in Figure 7.4 as discussed in Chapter 6 is given as [19]:

$$e_d = -\omega L i_q + L\frac{di_d}{dt} + (R + R_{SDBR})i_d + 0.5U_{dc}\beta_d \tag{7.14}$$

$$e_q = -\omega L i_d + L\frac{di_q}{dt} + (R + R_{SDBR})i_q + 0.5U_{dc}\beta_q \tag{7.15}$$

$$C\frac{dU_{dc}}{dt} = 0.75\left(i_d\beta_d + i_q\beta_q\right) - \frac{U_{dc}}{R_L} \tag{7.16}$$

$$r = \sqrt{\beta_d^2 + \beta_q^2} \tag{7.17}$$

From Equations (7.14) to (7.17), i_d, i_q represent the dq current input of the rectifier's axes, e_d, e_q are known as the dq voltage of the grid voltage axes components, ω is the angular frequency voltage, β_d, β_q represent the modulating signal of the rectifier's d and q axes components, while R_L, is the modulation signal vector norm. Considering three-phase transformation based on Park's principle, and the phase-A grid voltage is in alignment with the dq reference synchronous frame, the source voltage dq components are given as:

$$e_d = E_m \tag{7.18}$$

$$e_q = 0 \tag{7.19}$$

From Equation (7.18), U_{dc}, is the phase grid voltage amplitude. Consequently, the fed active and reactive rectifier's powers are computed by

$$P_s = \frac{3}{2}E_m i_d \tag{7.20}$$

$$Q_s = -\frac{3}{2}E_m i_q \tag{7.21}$$

To achieve power factor in unity mode of operation, i_{q_ref} can be set to 0. Therefore, for the current regulation to be ideal, $i_q = i_{qref} = 0$. With $i_q = 0$ and $e_q = 0$, the mathematical model of the VSC under unity power factor can be expressed with the following set of equations:

$$E_m = L\frac{di_d}{dt} + \left(R + R_{SDBR}\right)i_d + 0.5U_{dc}\beta_d \tag{7.22}$$

$$\beta_q = -\frac{2\omega L}{U_{dc}}i_q \tag{7.23}$$

$$C\frac{dU_{dc}}{dt} = \frac{3}{4}i_d\beta_d - \frac{U_{dc}}{R_L} \tag{7.24}$$

Equation (7.23) implies in order to ensure operation of unity power factor of the VSC, the component of β_q should proportionally vary with i_q current. From Equations (7.22) and (7.24), the capacity charge is manipulated by β_d, via the i_d current of the input.

The insertion of the SDBR resistance during transient state in the PMSG converter would affect the maximal power flow and the DC output voltage. During normal condition, the derivative operator relates all terms in Equations (7.22)–(7.24) would be zero. Thus, the new set of equations would be:

$$E_{\mathrm{m}} = (R + R_{\mathrm{SDBR}})i_{\mathrm{d}} + 0.5U_{\mathrm{dc}}\beta_{\mathrm{d}} \tag{7.25}$$

$$\beta_{\mathrm{q}} = -\frac{2\omega L}{U_{\mathrm{dc}}}i_{\mathrm{q}} \tag{7.26}$$

$$i_{\mathrm{d}} = \frac{4U_{\mathrm{dc}}}{3\beta_{\mathrm{d}}R_{\mathrm{L}}} \tag{7.27}$$

Putting Equation (7.26) into (7.27), for a given load of R_L, and voltage U_{dc}, the expression of the signal command β_{d} is

$$6E_m R_L \beta_{\mathrm{d}} - 8(R + R_{\mathrm{SDBR}})U_{dc} - 3R_L \beta_{\mathrm{d}}^2 U_{dc} = 0 \text{ for } \beta_{\mathrm{d}} \neq 0 \tag{7.28}$$

There are two derived solutions from Equation (7.28):

$$\beta_{\mathrm{d1}} = \frac{E_{\mathrm{m}}}{U_{\mathrm{dc}}} - \sqrt{\left(\frac{E_{\mathrm{m}}}{U_{\mathrm{dc}}}\right)^2 - \frac{8(R + R_{\mathrm{SDBR}})}{3R_L}} \tag{7.29}$$

$$\beta_{\mathrm{d2}} = \frac{E_{\mathrm{m}}}{U_{\mathrm{dc}}} + \sqrt{\left(\frac{E_{\mathrm{m}}}{U_{\mathrm{dc}}}\right)^2 - \frac{8(R + R_{\mathrm{SDBR}})}{3R_L}} \tag{7.30}$$

Since β_{d} component in Equation (7.29) has very low values, this solution is not admissible. Therefore, the solution of Equation (7.30) is the acceptable solution, making $\beta_{\mathrm{d}} = \beta_{\mathrm{d2}}$. And β_{d} would exist based on the following condition been satisfied:

$$\left(\frac{E_{\mathrm{m}}}{U_{\mathrm{dc}}}\right)^2 - \frac{8(R + R_{\mathrm{SDBR}})}{3R_{\mathrm{L}}} \geq 0 \tag{7.31}$$

The above condition is with respect to power consumption by the load. The operation of the converter is possible if

$$P_{\mathrm{dc}} \leq P_{\mathrm{dc\_max}} \tag{7.32}$$

where $P_{\mathrm{dc\_max}}$ is the maximal available power at the DC-connected side power of the PMSG converter. Considering the power conservation principle, $P_{\mathrm{dc\_max}}$ can be expressed as

$$P_{\mathrm{dc}} = \frac{3}{2}E_m i_{\mathrm{d}} - \frac{3}{2}(R + R_{\mathrm{SDBR}})i_{\mathrm{d}}^2 \tag{7.33}$$

By solving $dP_{dc} / di_d = 0$, the maximum power transfer to the DC side is

$$\frac{dP_{dc}}{di_d} = \frac{3}{2} E_m - 3Ri_d = 0 \rightarrow i_d = i_{d\_max} = \frac{E_m}{2R} \tag{7.34}$$

Putting Equation (7.34) into (7.33) yields the maximum DC side power as,

$$P_{dc\_max} = \frac{3E_m^2}{8(R + R_{SDBR})} \tag{7.35}$$

The PMSG VSC operation is possible when

$$P_{dc} \leq P_{dc\_max} \rightarrow \left(\frac{E_m}{U_{dc}}\right)^2 - \frac{8(R + R_{SDBR})}{3R_L} \geq 0 \tag{7.36}$$

Equation (7.36) reflects the condition of Equation (7.31). Finally, the grid input maximal power P_{s_max} can be obtained by putting Equation (7.35) into (7.21) for Q_s. Thus:

$$P_{s\_max} = \frac{3E_m^2}{4(R + R_{SDBR})} \tag{7.37}$$

Based on Equation (7.37), the maximum power transfer of the GSC of the PMSG, during transient state would be reduced by the insertion of SDBR in the GSC. Therefore, the total current reduces its value, thus, mitigating the oscillations that normally occur during transient conditions.

7.5 RESULTS AND DISCUSSIONS

Rigorous simulation studies were conducted to compare the fault ride through features in different grid conditions. The ratings of the parameters of both wind turbines are given in Table 7.1. The system performance was evaluated using PSCAD/EMTDC [20] environment. The fault types are symmetrical three-phase and unsymmetrical single-line-to-ground of 100 ms happening at 10.1 s, with the circuit breakers operation sequence opening and reclosing at 10.2 s and 11 s, respectively, on the faulted line at the terminals of both wind turbines. The fault performance with and without stability augmentation tools like SDBR and Overvoltage Protection System (OVPS) are presented below in detail.

7.5.1 OPERATION OF THE DFIG AND PMSG WIND TURBINES AT DIFFERENT GRID STRENGTHS

The DFIG and PMSG wind turbines were subjected to the weak grid, normal grid and strong grid, as shown in cases 1 to 3 in Table 7.2, considering no insertion of

TABLE 7.1

Rating of the Parameters of the Wind Turbines

DFIG wind turbine		PMSG wind turbine	
Rated power	5.0 MW	Rated power	5.0 MW
Stator resistance	0.01 pu	Stator resistance	0.01 pu
d-axis reactance	1.0 pu	d-axis reactance	1.0 pu
q-axis reactance	0.7 pu	q-axis reactance	0.7 pu
Machine inertia (H)	3.0	Machine inertia (H)	3.0
Effective DC-link protection	0.2 Ω	Effective DC-link protection	0.2 Ω
Effective SDBR	0.01 pu	Effective SDBR	0.05 pu
Overvoltage Protection System (OVPS)	110%	Overvoltage Protection System (OVPS)	110%

TABLE 7.2

Wind Turbines Operation at Different Grid Strengths

Cases	Grid strength	SCR	DFIG	PMSG
1	Weak	4	No SDBR	No SDBR
			No OVPS	No OVPS
2	Normal	8	No SDBR	No SDBR
			No OVPS	No OVPS
3	Strong	12	No SDBR	No SDBR
			No OVPS	No OVPS
4	Weak	4	With effective SDBR	With effective SDBR
			No OVPS	No OVPS
5	Weak	4	No SDBR	No SDBR
			With effective OVPS	With effective OVPS
6	Weak	4	With effective SDBR	With effective SDBR
			With effective OVPS	With effective OVPS
7	Weak	4	With 50% effective SDBR and OVPS	With 50% effective SDBR and OVPS

the SDBR and no OVPS. Some of the simulation results for the cases considered are shown in Figures 7.7–7.11.

From Figure 7.7, the overshoot of the DFIG and PMSG DC-link voltage variable could lead to the damage of the fragile and vulnerable power converters of the wind turbines during transient state. While the PMSG wind turbine experienced more overshoot and more settling time, the DFIG wind turbine has more DC-link voltage dip, with faster settling time. The overshoot and Dc-link voltage is more for case 1 (weak grid), compared to the other cases. The responses for the terminal voltage and the active power are shown in Figures 7.8 and 7.9, respectively. From Figures 7.9 and 7.10, more oscillations were observed for the active and reactive power variables

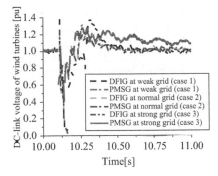

FIGURE 7.7   DC-link voltage of wind turbines at different grid strengths (cases 1,2,3).

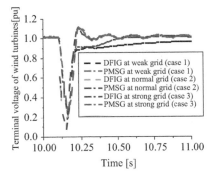

FIGURE 7.8   Terminal voltage of wind turbines at different grid strengths (cases 1, 2, 3).

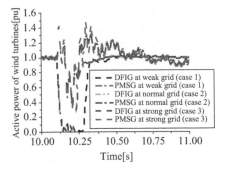

FIGURE 7.9   Active power of wind turbines at different grid strengths (cases 1, 2, 3).

with more oscillations and lower settling time during weak grid periods. However, the effect of the weak grid on the rotor speed of the DFIG wind turbine is negligible compared to the minimal effect observed in the PMSG wind turbine (Figure 7.11) during transient state for weak power grids. The effect of the power grid strengths is more obvious for the PMSG than the DFIG because the PMSG-based wind turbine is decoupled fully from the network due to the back-to-back power converter.

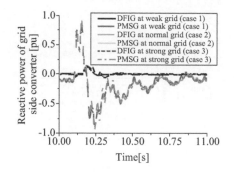

FIGURE 7.10   Reactive power of wind turbines at different grid strengths (cases 1, 2, 3).

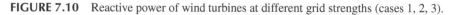

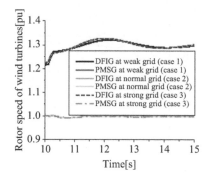

FIGURE 7.11   Rotor speed of wind turbines at different grid strengths (cases 1, 2, 3).

Consequently, the operation of the wind turbine in weak grids of case 1, is a critical situation during faulty condition, based on the presented simulation results in Figures 7.7–7.11. In light of this, the subsequent sections of this chapter would consider the improvement of the wind turbine's fault ride through in weak grids considering the employment of SDBR and OVPS schemes.

7.5.2   IMPROVING THE PERFORMANCE OF **DFIG** AND **PMSG** WIND TURBINES IN WEAK GRIDS CONSIDERING THE EFFECTIVE SIZING OF **SDBR**

The performances of the DFIG and PMSG wind turbines in weak grids were further enhanced considering the use of an effective-sized SDBR shown in Table 7.1. The SDBR was connected at the stator side of both wind turbines for effective comparison during transient state. In this considered case summarized in Table 7.2 (cases 1 and 4), no overvoltage protection scheme was considered. Some of the simulation results for the key variables of the wind turbines are shown in Figures 7.12–7.14. The undershoot, overshoot and settling time of the DC-link voltage, terminal voltage and active power of the DFIG and PMSG wind turbine variables without the SDBR connected at the stator was more, as shown in Figures 7.12–7.14. However, placing the SDBR on GSC of both wind turbines has great influence on the wind

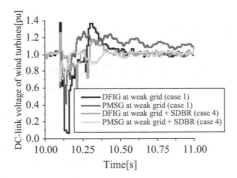

FIGURE 7.12 DC-link voltage of wind turbines with SDBR at weak grid (cases 1 and 4).

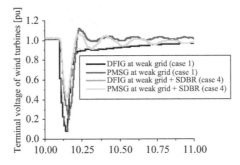

FIGURE 7.13 Terminal voltage of wind turbines with SDBR at weak grid (cases 1 and 4).

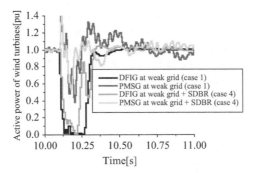

FIGURE 7.14 Active power of wind turbines with SDBR at weak grid (cases 1 and 4).

generator variables in fault conditions. This is because the expected wind generator stator circuitry of the DFIG and PMSG high voltage is divided by the SDBR since its connection is series topology. As seen from Figures 7.12–7.14, the undershoot, overshoot and time of settling for the wind generator variables were much improved considering the effective-sized SDBR. Consequently, there is no loss in control of the power converters, since no induced overvoltage would be experienced. More so, during operation, the SDBR has the capability to mitigate high current flow significantly, thus, avoiding dangerous overvoltage and high charging current normally experienced in power converters that are vulnerable and fragile in nature.

7.5.3　Improving the Performance of DFIG and PMSG Wind Turbines in Weak Grids Considering Overvoltage Protection System (OVPS)

The performances of the DFIG and PMSG wind turbines in weak grids were further enhanced considering the use of an effective-sized OVPS shown in Table 7.1. The OVPS is a DC chopper connected at the DC-link of both wind turbines for effective comparison during transient state. In this considered case summarized in Table 7.2 (cases 1 and 5), no SDBR scheme was considered. The DC-link voltage variable of the wind turbines are shown in Figure 7.15. The undershoot, overshoot and settling time of the DC-link voltage of the DFIG and PMSG wind turbine variable without considering OVPS were more than when OVPS was considered. This would protect the power converters of the wind turbines because the wind turbines power converters would be operating within its permissible limits.

7.5.4　Improving the Performance of DFIG and PMSG Wind Turbines in Weak Grids Considering SDBR and Overvoltage Protection System

The performances of the DFIG and PMSG wind turbines in weak grids were further enhanced considering the combination of an effective-sized SDBR and OVPS shown in Table 7.1. In this considered case summarized in Table 7.2 (cases 1 and 6), a combination of SDBR and OVPS schemes would definitely improve the performance of the wind generator variables, as shown in the DC-link voltage and terminal voltage variables of the wind turbines in Figures 7.16 and 7.17.

The undershoot, overshoot and settling time of the DC-link voltage, terminal voltage of the DFIG and PMSG wind turbines without considering SDBR and OVPS in weak grids were more as shown in Figures 7.16 and 7.17, respectively. However, in order to improve the fault ride through of the wind turbines, a combination of the SDBR and OVPS would protect the power converters and enhance the variables of

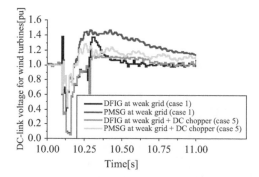

FIGURE 7.15　DC-link voltage of wind turbines with OVPS at weak grid (cases 1 and 5).

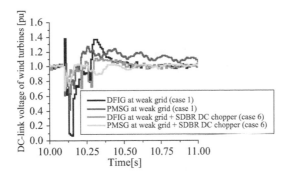

FIGURE 7.16 DC-link voltage of wind turbines with SDBR and OVPS at weak grid (cases 1 and 6).

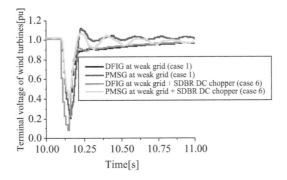

FIGURE 7.17 Terminal voltage of wind turbines with SDBR and OVPS at weak grid (cases 1 and 6).

the wind turbines. Thus, this type of hybrid fault ride through technique is recommended for variable speed wind turbines operating at weak power grids.

7.5.5 IMPROVING THE PERFORMANCE OF DFIG AND PMSG WIND TURBINES IN WEAK GRIDS CONSIDERING 75 AND 50% OF THE EFFECTIVE SDBR AND OVERVOLTAGE PROTECTION SYSTEM

The performances of the DFIG and PMSG wind turbines in weak grids were further investigated considering the combination of 75 and 50% effective-sized SDBR and OVPS shown in Table 7.1. In this considered case summarized in Table 7.2 (cases 1 and 7), the proposed hybrid scheme is proven to enhance the performance of the wind turbines at transient state even with reduced value of 75 and 50% of their effective values, as shown in Figures 7.18 and 7.19 for 75% reduction and Figures 7.20 and 7.21 for 50% reduction, for the DC-link voltage and the terminal voltage of the wind turbines. Thus, it is better to use half of the effective values of the combined schemes, than using either SDBR or OVPS.

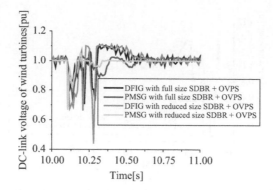

FIGURE 7.18 DC-link voltage of wind turbines with 75% SDBR and OVPS at weak grid (cases 1 and 7).

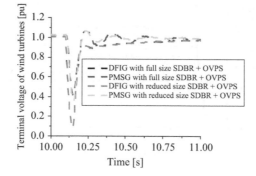

FIGURE 7.19 Terminal voltage of wind turbines with 75% SDBR and OVPS at weak grid (cases 1 and 7).

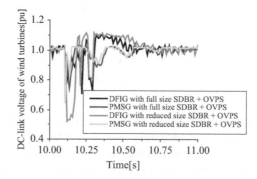

FIGURE 7.20 DC-link voltage of wind turbines with 50% SDBR and OVPS at weak grid (cases 1 and 7).

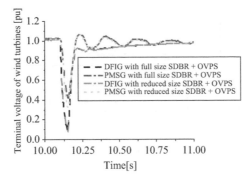

FIGURE 7.21 Terminal voltage of wind turbines with 50% SDBR and OVPS at weak grid (cases 1 and 7).

7.5.6 Investigating the Performance of DFIG and PMSG Wind Turbines in Weak Grids Considering the Effective SDBR and Overvoltage Protection System in a Single-Line-to-Ground Fault

The study of the performance of the DFIG and PMSG wind turbines in weak grids was extended for the single-line-to-ground fault, which is an example of unsymmetrical fault as shown in Figures 7.22 and 7.23, respectively. The combination of an effective-sized SDBR and OVPS shown in Table 7.1, were considered, as summarized in Table 7.2 (cases 1 and 6). From the results in Figures 7.22 and 7.23, a combination of SDBR and OVPS schemes would definitely improve the performance of the wind generator variables even during unsymmetrical fault conditions.

For 50% reduction of the effective SDBR and OVPS during a single-line-to-ground unsymmetrical fault, the response for the DC-link voltage and the terminal voltage of the wind turbines are shown in Figures 7.24 and 7.25, respectively. From the obtained results, is better to use half of the effective values of the combined schemes, than using either SDBR or OVPS, as obtained earlier using a severe three-phase-to-ground fault.

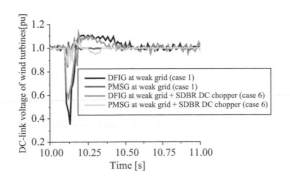

FIGURES 7.22 DC-link voltage of wind turbines with SDBR and OVPS at weak grid (cases 1 and 6).

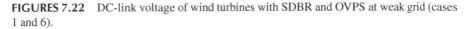

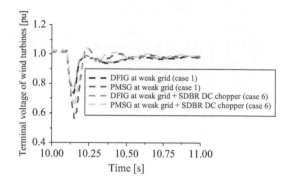

FIGURE 7.23   Terminal voltage of wind turbines with SDBR and OVPS at weak grid (cases 1 and 6).

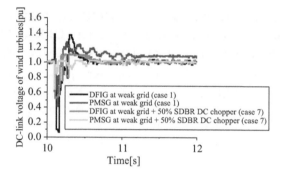

FIGURE 7.24   DC-link voltage of wind turbines with 50% SDBR and OVPS at weak grid (cases 1 and 7).

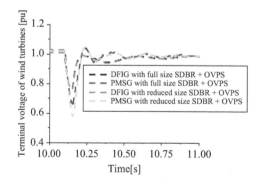

FIGURE 7.25   Terminal voltage of wind turbines with 50% SDBR and OVPS at weak grid (cases 1 and 7).

As seen from the mathematical dynamics of the SDBR in the wind turbines presented in this chapter, the SDBR is closely related to the stator resistance of the wind turbines. Also, the maximum power transfer of the GSC of the wind turbines, during transient state would be reduced by the insertion of SDBR in the GSC. Therefore, by reducing the percentage of the SDBR or OVPS, the maximum power transfer would be slightly increased. This means more rating of the wind generator would be required, while keeping the performance satisfactory.

7.6   CHAPTER CONCLUSION

This chapter presented a comparative evaluation of the performance of DFIG and PMSG wind turbines with the same machine ratings. The wind turbines were operated at weak, normal and strong grids to show the influence of the network parameters on the wind turbines. Both wind turbines were affected by the various strengths of the power grid. However, the influence of the power grid strengths was more observed in the PMSG wind turbine compared to the DFIG wind turbine. This is because the PMSG wind turbine is decoupled at the grid or network side by its full back-to-back power converters. The weak power grids would result in critical situation of operation during faulty conditions, compared to the normal and strong power grids. Consequently, further analysis using the weak power grid was considered in this chapter for both wind turbines, considering the use of SDBR and OVPS. The hybrid scheme of the SDBR and OVPS in both wind turbines was able to improve the performance of the variables of the wind turbine and keep the operation of the power converters within their permissible limits. It was observed that, even if a 50% reduction in SDBR or OVPS, the performance is still satisfactory, as shown in Section 7.5.5 of this chapter.

Therefore, it is recommended to use the combination of the SDBR and OVPS with DFIG or PMSG-based variable speed wind turbines to get superior fault ride through performance, especially in weak grid conditions.

REFERENCES

[1] N. P. W. Strachan, and D. Jovcic, "Stability of a variable-speed permanent magnet wind generator with weak ac grids," *IEEE Transactions on Power Delivery*, vol. 25, no. 4, pp. 2779–2788, 2010.

[2] J. Hu, Y. Huang, D. Wang, H. Yuan, and X. Yuan, "Modeling of grid-connected DFIG-based wind turbines for dc-link voltage stability analysis," *IEEE Transactions on Sustainable Energy*, vol. 6, no. 4, pp. 1325–1336, 2015.

[3] Y. Zhou, D. D. Nguyen, P. C. Kjr, and S. Saylors, "Connecting wind power plant with weak grid - challenges and solutions," *2013 IEEE Power Energy Society General Meeting*, pp. 1–7, 2013.

[4] M. Garcia-Garcia, M. P. Comech, J. Sallan and A. Liombart, "Modelling wind farms for grid disturbances studies," *Science Direct, Renewable Energy*, vol. 33, pp. 2019–2121, 2008.

[5] K. E. Okedu, S. M. Muyeen, R. Takahashi, and J. Tamura, "Application of SDBR with DFIG to augment wind farm fault ride through," *24th IEEE-ICEMS (International Conference on Electrical Machines and Systems)*, August 2011, Beijing, China, pp. 1–6.

[6] K. E. Okedu, "Enhancing the performance of DFIG variable speed wind turbine using parallel integrated capacitor and modified modulated braking resistor," *IET Generation Transmission & Distribution*, vol. 13, no. 15, pp. 3378–3387, 2019.

[7] K. E. Okedu, "Improving the transient performance of DFIG wind turbine using pitch angle controller low pass filter timing and network side connected damper circuitry," *IET Renewable Power Generation*, vol. 14, no. 7, pp. 1219–1227, 2020.

[8] I. Zubia, J. X. Ostolaza, A. Susperrgui, and J. J. Ugartemendia, "Multi-machine transient modeling of wind farms, an essential approach to the study of fault conditions in the distribution network," *Applied Energy*, 89, 1, pp. 421–429, 2012.

[9] X. Kong, Z. Z. Xianggen, and M. Wen, "Study of fault current characteristics of the DFIG considering dynamic response of the RSC," *IEEE Transactions Energy Conversion*, 2, 2, pp. 278–287, 2014.

[10] Y. Li, Z. Xu, and K. P. Wong, "Advanced control strategies of PMSG-based wind turbines for system inertia support," *IEEE Transactions on Power Systems*, vol. 32, pp. 3027–3037, 2017.

[11] N. Priyadarshi, V. Ramachandaramurthy, S. Padmanaban, and F. Azam, "An ant colony optimized MPPT for standalone hybrid PV-wind power system with single Cuk converter," *Energies*, vol. 12, p. 167, 2019.

[12] S. W. Lee, and K. H. Chun, "Adaptive sliding mode control for PMSG wind turbine systems," *Energies*, vol. 12, pp. 595, 2019.

[13] S. Heier, "Wind energy conversion systems," In *Grid Integration of Wind Energy Conversion Systems*, Chicester, UK: John Wiley & Sons Ltd, 1998, pp. 34–36.

[14] MathWorks, MATLAB documentation center. http://www.mathworks.co.jp/jp/help/ (accessed on 12 March 2012).

[15] S. M. Muyeen, A. Al-Durra, and J. Tamura, "Variable speed wind turbine generator system with current controlled voltage source inverter," *Energy Conversion and Management*, vol. 52, no. 7, pp. 2688–2694, 2011.

[16] S. Li, T. A. Haskew, and L. Xu, "Conventional and novel control design for direct driven PMSG wind turbines," *Electric Power Systems Research*, 80, 328–338, 2010.

[17] D. Yang, X. Ruan, and H. Wu "Impedance shaping of the grid-connected inverter with LCL filter to improve its adaptability to the weak grid condition," *IEEE Transactions on Power Electronics*, vol. 29, no. 11, pp. 5795–5805, 2014.

[18] C. V. Thierry. *Voltage Stability of Electric Power Systems*. New York, US: Springer, 1998.

[19] S. Grunau, and W. F. Fuchs. "Effect of wind-energy power injection into weak grids," *Institute for Power Electronics and Electrical Drives, Christian-Albrechts-University of Kiel D-24143*, Kiel, Germany, pp. 1–7, 2012.

[20] PSCAD/EMTDC Manual, Manitoba HVDC lab, 2016.

8 DFIG Wind Turbines and Supercapacitor Scheme

8.1 CHAPTER INTRODUCTION

The energy storage systems (ESS) model is one of the most widely used control strategies in wind energy conversion system. One of the ESS elements is a supercapacitor that enables smooth output power, with the ability of mitigating oscillations. Constant power control could be provided in DFIG wind turbines by using the supercapacitor model. In the literature, a management control unit has been developed to provide power tracking in a two-layer control scheme in [1] and constant power control in [2, 3], considering the supercapacitor model. In [4], a study on the suppression of power fluctuation in DFIG based on supercapacitor energy storage was carried out, in order to improve its performance. The study elaborated that supercapacitors could be used in DFIGs to obtain constant output power as well as to prevent frequency changes of switching elements in the converter circuit. The study monitored the maximum power point of a DFIG wind turbine, considering the supercapacitors scheme. In another study carried out in [5, 6], the responses to the change in wind speed, tower shadow and protection units were investigated. The supercapacitor model was selected in order to optimize energy production and consumption by adjusting the output power in the DFIG. The safe and effective use of the supercapacitors was observed under optimal operating conditions in [7, 8], for a battery hybrid energy storage system and wind energy integration. In addition to ensuring optimum operating conditions, a coordinated control was carried out in the DFIG wind turbine considering the supercapacitor scheme in two stages for optimal active power control management [9]. The supercapacitor was selected in order to provide pitch angle control in the DFIG because of its advantages which include high performance, high temperature operating ability, long service life and convenient use in applications [10].

In this chapter, a new control strategy of a supercapacitor system is examined for a DFIG wind turbine during severe grid fault condition. A simple two-machine model system consisting of a DFIG connected to a fixed speed squirrel cage induction generator wind turbine was used in carrying out the transient analysis in this study. The supercapacitor system was connected across the terminals of the DC-link voltage of the DFIG wind turbine, between the RSC and the GSC. The study considered the determination of the effective parameters of the supercapacitor in the DFIG wind turbine, by considering different values of resistance, inductance and capacitance of the supercapacitor, in various scenarios. Furthermore, two switching strategies of the embedded supercapacitor system in the DFIG, the DC-link voltage and the grid voltage, were investigated using the various values earlier considered. The obtained results were compared with those using parallel capacitor scheme for the DFIG connected to the RSC and GSC of the wind turbine. For effective comparative study,

DOI: 10.1201/9781003350910-8

between the supercapacitor and parallel capacitor-based solution to buffer the transient energy of the DFIG, the same capacitance value was employed in the study. The proposed supercapacitor scheme for the DFIG wind turbine has the ability to store a large amount of electric charge compared to parallel capacitors and all other types of conventional capacitors. This is because the charge storage, which is the capacitance in conventional capacitors, is directly proportional to the surface area of each electrode or plate and inversely proportional to the distance between them. The topology of capacitors and batteries in wind turbines differs in two ways considering the amount of charge or energy storage and how quickly the charge or energy is dissipated. The batteries have the ability to store more energy than conventional capacitors, however, the shortcoming of the batteries is that they cannot deliver the stored energy very quickly. Though the capacitors deliver the stored energy more quickly, however, they lack the ability to store a large amount of energy as the batteries. These two shortcomings could be overcome by using the proposed supercapacitors in DFIG wind turbines. This is because the supercapacitors have large surface area electrodes and very thin dielectric, as separating distance between the electrodes. Thus, compared to the conventional capacitors, the supercapacitors have very small distance between the electrodes, making it possible to obtain larger capacitance or energy storage that is delivered quickly to help improve the performance of the DFIG wind turbine variables during transient state [11].

8.2 MODELING AND CONTROL

8.2.1 Wind Turbine Characteristics and DFIG Control

The DFIG wind turbine characteristics, modeling and control could be referred to in the previous chapters of the book (Chapters 1–3, 5 and 7).

8.3 THE DFIG MODEL WITH SUPERCAPACITOR SYSTEM

8.3.1 The Dynamics of the Traditional Supercapacitor System

The traditional simple electrical equivalent model of the supercapacitor is shown in Figure 8.1. The model cell capacity expression is given by [12, 13]:

$$C_{cell} = C_0 + K_v V_c \tag{8.1}$$

The total capacity expressions for n numbers in the series for the model system are:

$$C_{total} = \cfrac{1}{\cfrac{1}{C_{cell1}} + \cfrac{1}{C_{cell2}} + \cfrac{1}{C_{cell3}} + \dots \cfrac{1}{C_{celln}}} \tag{8.2}$$

$$C_{total} = \frac{1}{n} C_{cell} = \frac{1}{n}\left(K_v V_c\right) \tag{8.3}$$

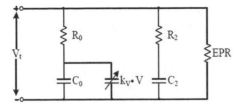

FIGURE 8.1 Traditional supercapacitor model.

The model terminal voltage equations and the capacity chance over time are expressed as:

$$V(t) = i(t)R + \frac{1}{C_{total}} \int i(t)dt \tag{8.4}$$

$$V(t) = V_C(t) + (C_0 + K_vV_c)(R_0 + R_2) + \frac{dV_c(t)}{dt} \tag{8.5}$$

$$\frac{dV_c}{dt} = \frac{V - V_c}{(R_0 + R_2)(C_0 + K_vV_c)} \tag{8.6}$$

In the supercapacitor model in Figure 8.1, depending on the number of cells, the use of multiple resistors can be represented by Equivalent Series Resistance (ESR) and Equivalent Parallel Resistance (EPR). The voltage and initial voltage equations using the equivalent series resistors are given by:

$$V(t) = R_{ESR}i(t) + \frac{1}{C_{total}} \int i(t)dt \tag{8.7}$$

$$V_0 = R_{ESR}i(t) + \frac{1}{C_{total}} \int i(t)dt \tag{8.8}$$

$$R_{ESR} + C_{total}\frac{di(t)}{dt} - i(t) = 0 \tag{8.9}$$

The charge and discharge expressions in the supercapacitor system are given in Equations (8.10) and (8.11), respectively, while the expression for the terminal voltage as a function of time is given in Equation (8.12).

$$V_r(t) = Ke^{\frac{1}{R_{ESR} + C_{total}}t} \tag{8.10}$$

$$\frac{dV_c}{dt} = \frac{-V_c}{(R_0 + R_2)(C_0 + K_v V_c)}$$

(8.11)

$$V_t(t) = V_c(t) + V_r(t)$$

(8.12)

8.3.2 THE DYNAMICS OF THE SUPERCAPACITOR SYSTEM IN DFIG WIND TURBINES

The connection of the supercapacitor in the DFIG is shown in Figure 8.2, where P is the grid side transformer power, P_{grid} is the power of the grid, P_s is the stator power, P_r is the rotor power and $P_{supercapacitor}$ is the power of the supercapacitor. In the DFIG, the supercapacitor is able to adjust the DC bus voltage value in the range of 0–100%. While certain parts of the power values are met by the grid in the creation of the supercapacitor model, the remaining power values are met by the DFIG. The amount of energy and capacity expressions stored in the supercapacitor are given in Equations (8.13)–(8.15).

$$E_{supercapacitor} = 0.2 P_{nominal} t$$

(8.13)

$$E_{supercapacitor} = \frac{1}{2} C_{supercapacitor} \left(V_{max}^2 - V_{min}^2 \right)$$

(8.14)

$$C_{supercapacitor} = \frac{0.4 P_{nominal} t}{V_{max}^2 - V_{min}^2}$$

(8.15)

where $E_{supercapacitor}$ is the amount of energy in the supercapacitor, $P_{nominal}$ is the nominal power value, t is the supercapacitor operating time, $C_{supercapacitor}$ is the supercapacitor capacity value, V_{max} is the maximum supercapacitor voltage and V_{min} is the minimum supercapacitor voltage, respectively.

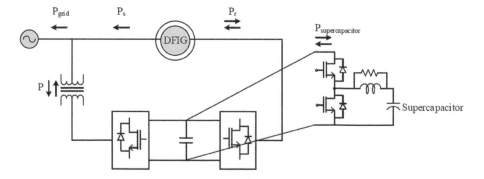

FIGURE 8.2   DFIG with the proposed supercapacitor topology [14].

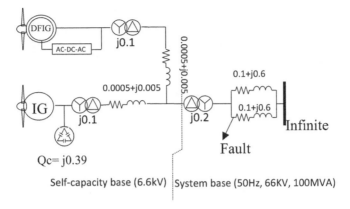

(a)

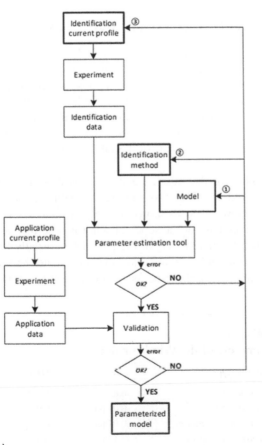

(b)

FIGURE 8.3 (a) Model system of study. (b) Parameter estimation procedure. (*Continued*)

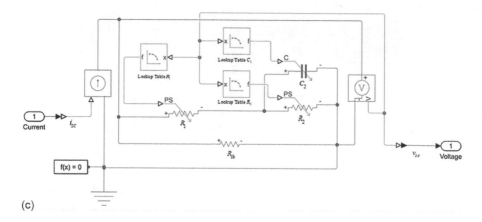

(c)

FIGURE 8.3 (Continued)   (c) Simscape model.

8.4   THE MODEL SYSTEM OF STUDY AND PARAMETERS

The model system used for this study is shown in Figure 8.3(a) and the related parameters of the wind turbines are given in Table 8.1. The excitation parameters of the DFIG are given in Table 8.2, while the parameters of the supercapacitor system for the different cases considered are given in Table 8.3. In the model system of Figure 8.3(a), the DFIG and Induction Generator (IG) wind turbines were connected to an infinite bus bar and subjected to a severe three-phase-to-ground fault. The supercapacitor was connected to the terminals of the DFIG wind turbine in Figure 8.3, as shown in Figure 8.2.

The switching strategy of the supercapacitor is based on the DC-link voltage exceeding the set threshold of 110% of its nominal value during transient state or the grid voltage dropping below 1.0 pu, as shown in Tables 8.2 and 8.3, respectively. The parameter estimation procedure for the supercapacitor is described as follows based on Figure 8.3(b). Generally, the models of the parameters usually have differential equations, transfer functions or block diagrams, that are updated offline or online. To obtain offline mode parameters, the process involves storing the data to use them

TABLE 8.1
Parameters of the Wind Turbines

Generator type	IG	DFIG
Rated voltage	690 V	690 V
Stator resistance	0.01 pu	0.01 pu
Stator leakage reactance	0.07 pu	0.15 pu
Magnetizing reactance	4.1 pu	3.5 pu
Rotor resistance	0.007 pu	0.01 pu
Rotor leakage reactance	0.07 pu	0.15 pu
Inertia constant	1.5 s	1.5 s

TABLE 8.2

Excitation Parameters and Switching Threshold of the DFIG Wind Turbine

DC-link voltage	1.5 kV
DC-link capacitor	50,000 μF
Device for power converter	IGBT
PWM carrier frequency	2 kHz
Upper limit of DC voltage switching (E_{dc_Max})	1.65 kV (110%)
Lower limit of DC voltage switching (E_{dc_Min})	0.75 kV (50%)
Short circuit parameter of protective device for overvoltage	0.2 Ohm
Grid voltage	≥ 1.0 pu Normal condition
	< 1.0 pu Faulty condition

TABLE 8.3

Parameters and Switching Strategies of the Supercapacitor

Case	DC-link voltage switching strategy			Grid voltage switching strategy		
	R (Ω)	L (H)	C (F)	R (Ω)	L (H)	C (F)
1	0.1	1	1	0.1	1	1
2	0.2	2	2	0.2	2	2
3	0.3	3	3	0.3	3	3
4	0.1	1	1	0.1	1	1
5	0.2	2	2	0.2	2	2
6	0.3	3	3	0.3	3	3

much later, while for the online mode, it is based on a parallel experiment [15]. However, there exist many procedures to achieve supercapacitor parameters like an unscented Kalman filter [16] or the Luenberger-style scheme [17]. In this chapter, the supercapacitor parameters were selected based on interactive, simple and offline procedures [18] in Figure 8.3(b) considering the Simscape model of Figure 8.3(c), respectively.

8.5 SIMULATION RESULTS AND DISCUSSIONS

8.5.1 EVALUATION OF THE PROPOSED DFIG SUPERCAPACITOR SCHEME

Rigorous simulation studies were conducted to compare the fault ride through features of the DFIG supercapacitor-based system connected to a fixed speed induction generator wind turbine shown in Figure 8.3(a) model system. The system performance was evaluated using PSCAD/EMTDC [19] environment. The fault type is a severe symmetrical three-phase of 100 ms happening at 0.1 s, with the circuit breakers operation sequence opening and reclosing at 0.2 s and 1.0 s, respectively, on the faulted line at the fault point shown in the model system of Figure 8.3(a). The fault performance with different parameters and switching strategies of the stability augmentation tool of the supercapacitor is presented below in detail.

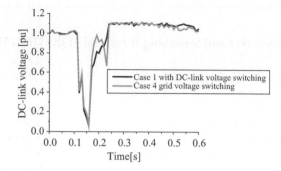

FIGURE 8.4 DC-link voltage (cases 1 and 4).

The DFIG wind turbine with supercapacitor scheme was subjected to cases 1–6 in Table 8.3, considering the excitation parameters and switching thresholds in Table 8.2. Some of the simulation results for the cases considered are shown in Figures 8.4–8.12. In Figures 8.4 and 8.5, the DC-link voltage was not able to recover on time after the grid fault using both switching strategies of DC-link and grid voltage for cases 1, 4, 2 and 5, respectively. However, for cases 3 and 6, in Figure 8.6, the performance of the DC-link voltage was the same for both switching strategies. Thus, the effective parameters of the supercapacitor for better performance of the DFIG wind turbine during transient state are 0.3Ω, 3H, 3F, for R, L, C, respectively, in Table 8.3. Figures 8.7–8.9 show the terminal voltage for the DFIG and IG wind turbines. From the figures, the parameters of the supercapacitors do not have effects on the responses of the terminal voltage of the wind turbines.

In Figure 8.10(a) and (b) the active power was more influenced in case 6, compared to the other cases using the supercapacitor scheme, while in Figure 8.11(a) and (b), the reactive power was also more dissipated or enhanced in case 6 compared to the other cases. It was also observed from the DFIG rotor speed performance in Figure 8.12(a) and (b) that the transient state performance in case 6 gave better response. Therefore, the effective parameters for the improved performance of the DFIG supercapacitor embedded system are when the resistance, inductance and

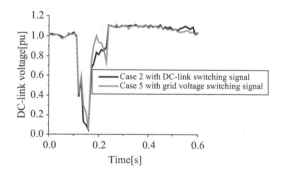

FIGURE 8.5 DC-link voltage (cases 2 and 5).

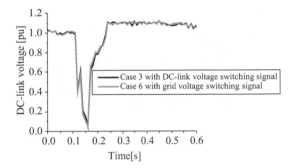

FIGURE 8.6 DC-link voltage (cases 3 and 6).

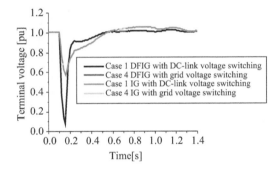

FIGURE 8.7 Terminal voltage of DFIG (cases 1 and 4).

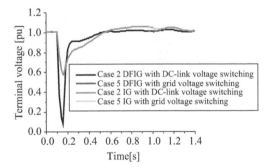

FIGURE 8.8 Terminal voltage of DFIG (cases 2 and 5).

capacitance values are not too small. This is because the terminal voltage of the generator increases, mitigating the depression of the electrical torque and power. The supercapacitor will increase the mechanical power extracted from the drive train, thus reducing its speed excursion. Also, since mechanical torque is proportional to the square of the stator voltage of the DFIG, the effect would enhance the post-fault recovery of the DFIG wind turbine.

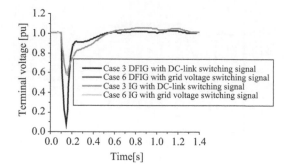

FIGURE 8.9   Terminal voltage of DFIG (cases 3 and 6).

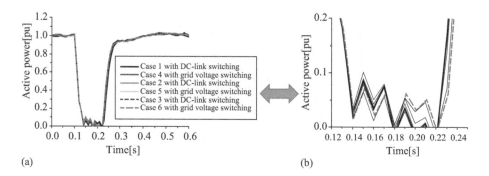

FIGURE 8.10   (a) Active power of DFIG for all cases. (b) Zoom of 8.10 (a).

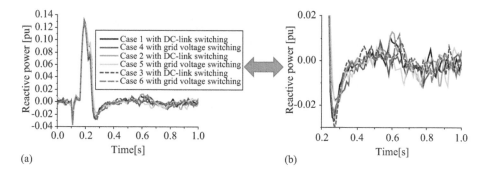

FIGURE 8.11   (a) Reactive power of DFIG for all cases. (b) Zoom of 8.11 (a).

8.5.2 EVALUATION OF THE PROPOSED DFIG SUPERCAPACITOR SCHEME AND PARALLEL DFIG CAPACITOR SCHEME

In this section of this chapter, the proposed supercapacitor scheme and parallel capacitor-based scheme with the same capacitance value was evaluated for the DFIG, considering Figure 8.13(a), with a conventional DC chopper circuit connected

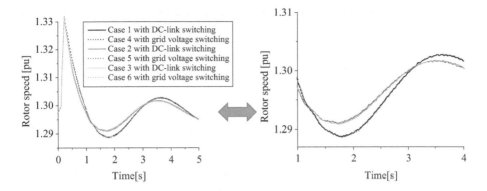

FIGURE 8.12 (a) Rotor speed of DFIG for all cases. (b) Zoom of 8.12 (a).

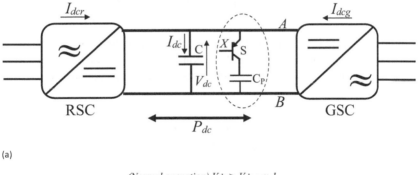

(a)

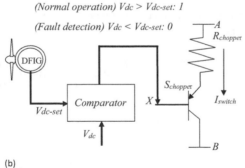

(b)

FIGURE 8.13 Parallel capacitor topology for DFIG wind turbine. (a) Parallel capacitor-based DFIG DC-link. (b) Switching scheme for chopper and parallel capacitor.

between the power converters of the wind turbine. The parallel capacitor scheme was connected at both the RSC and the GSC and the switching strategy for both connections is shown in Figure 8.13(b). Figure 8.13 shows the topology of the DFIG-based parallel capacitor scheme. The mathematical dynamics of connecting the parallel capacitor to the DFIG are given as follows.

As shown in Figure 8.13, the power flowing via the DC-link circuit can be expressed as [20, 21]:

$$P_{\text{converter}} = V_{\text{dc}}i_{\text{dcr}} = -V_{\text{dc}}i_{\text{dcg}} = -\frac{3}{2}v_{\text{gq}}i_{\text{gq}} \tag{8.16}$$

$$\left(C+C_{\text{p}}\right)\frac{dV_{\text{dc}}}{dt} = i_{\text{dcg}} + i_{\text{dcr}} \tag{8.17}$$

Putting the i_{dcg} term in Equation (8.17) with i_{gq}, the DC-link voltage and q-component current relationship can be found as follows:

$$\left(C+C_{\text{p}}\right)\frac{dV_{\text{dc}}}{dt} = \frac{3}{2}\frac{v_{\text{gq}}i_{\text{gq}}}{V_{\text{dc}}} + i_{\text{dcr}} \tag{8.18}$$

In Equation (8.18), the grid quantities are related to the first term, while the RSC injecting currents are associated with the second term. This rotor-injecting current is an input disturbance caused by the power change. As a result, Equation (8.18) can be re-written as

$$\left(C+C_{\text{p}}\right)\frac{dV_{\text{dc}}}{dt} = \frac{3}{2}\frac{v_{\text{gq}}i_{\text{gq}}}{V_{\text{dc}}} + \frac{P_{\text{converter}}}{V_{\text{dc}}} = f \tag{8.19}$$

If Equation (8.16) is differentiated with respect to all variables considering a given point v_{gq0}, i_{gq0}, V_{dc0}, then:

$$\left(C+C_{\text{p}}\right)\Delta\dot{V}_{\text{dc}} = \frac{\partial f}{\partial i_{\text{gq}}}\Delta i_{\text{gq}} + \frac{\partial f}{\partial v_{\text{gq}}}\Delta v_{\text{gq}} + \frac{\partial f}{\partial V_{\text{dc}}}\Delta V_{\text{dc}} + \frac{\partial f}{\partial i_{\text{dcr}}}\Delta i_{\text{dcr}} \tag{8.20}$$

$$\Rightarrow \Delta P_{\text{converter}} = V_{\text{dc0}}\Delta i_{\text{dcr}}$$

$$s\left(C+C_{\text{p}}\right)\Delta V_{\text{dc}} = \frac{3}{2}\frac{v_{\text{gq0}}}{V_{\text{dc0}}}\Delta i_{\text{gq}} + \frac{3}{2}\frac{i_{\text{gq0}}}{V_{\text{dc0}}}\Delta v_{\text{gq}} - \frac{3}{2}\frac{v_{\text{gq0}}i_{\text{gq0}}}{V_{\text{dc0}}^2}\Delta V_{\text{dc}} + \frac{\Delta P_{\text{converter}}}{V_{\text{dc0}}}$$

$$= \frac{3}{2}K_{\text{V}}\Delta i_{\text{gq}} + \frac{3}{2}K_{\text{G}}\Delta v_{\text{gq}} + \frac{1}{V_{\text{dc0}}}\Delta P_{\text{converter}} - \frac{3}{2}K_{\text{V}}K_{\text{G}}\Delta V_{\text{dc}} \tag{8.21}$$

From Equation (8.21), $K_{\text{V}} = \dfrac{v_{\text{gq0}}}{V_{\text{dc0}}}$ and $K_{\text{G}} = \dfrac{i_{\text{gq0}}}{V_{\text{dc0}}}$

Figures 8.14 and 8.15 show the comparative analysis of the proposed DFIG super-capacitor scheme and the conventional DFIG parallel capacitor scheme. In Figures 8.14

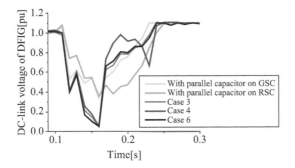

FIGURE 8.14 DC-link voltage of DFIG wind turbine.

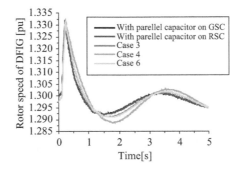

FIGURE 8.15 Rotor speed of DFIG wind turbine.

and 8.15, when the parallel capacitor was connected to the GSC of the DFIG power converter, better response was observed for the DC-link voltage and rotor speed of the wind generator, with fast recovery of the variables after transient state. The connection of the parallel converter at the RSC led to delayed recovery of the wind generator DC-link voltage and rotor speed variables. However, the proposed supercapacitor DFIG-based system for case 6 with optimal parameter ratings gave optimal changes than cases 3 and 4 and also the conventional DFIG parallel capacitor scheme.

8.5.3 EVALUATION OF THE PROPOSED DFIG SUPERCAPACITOR SYSTEM DURING ASYMMETRICAL FAULTS AT SUPER-SYNCHRONOUS AND SUB-SYNCHRONOUS SPEEDS

A further analysis of the performance of the proposed approach during super-synchronous as well as sub-synchronous speeds was carried out in this section, as this will significantly affect the LVRT capability of the DFIG during asymmetrical faults. Figures 8.16 and 8.17 show the responses of the DFIG wind turbine during super-synchronous speed, when the wind speed is above the nominal or rated wind speed, and the sub-synchronous speed, when the wind speed is below the nominal or rated wind speed. In Figures 8.16 and 8.17, the performances of the DFIG DC-link

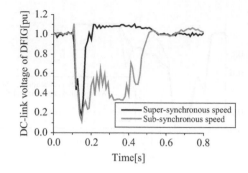

FIGURE 8.16   DC-link voltage of DFIG wind turbine 2LG.

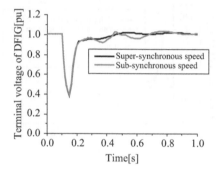

FIGURE 8.17   Terminal voltage of DFIG wind turbine 2LG.

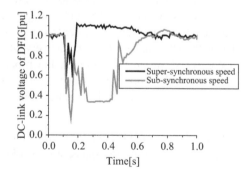

FIGURE 8.18   DC-link voltage of DFIG wind turbine 2LL.

voltage and terminal voltage were better during the super-synchronous speed than the sub-synchronous speed, during the two-line-to-ground fault scenario, because the wind generator is operating above its rated power during the fault scenario. Similarly, the same performance is expected for the line-to-line and line-to-ground faults in Figures 8.18–8.21, for the DC-link voltage and terminal voltage of the DFIG wind turbine.

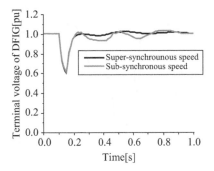

FIGURE 8.19 Terminal voltage of DFIG wind turbine 2LL.

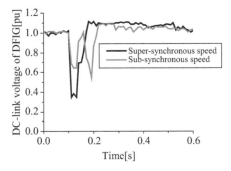

FIGURE 8.20 DC-link voltage of DFIG wind turbine 1LG.

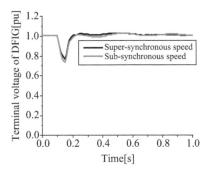

FIGURE 8.21 DC-link voltage of DFIG wind turbine 1LG.

8.5.4 PERFORMANCE OF THE PROPOSED SCHEME UNDER ZERO-VOLTAGE CONDITION AT THE TERMINAL OF THE MACHINE

In this section of this chapter, the performance of the proposed scheme was carried out under zero-voltage condition at the terminal of the DFIG wind turbine, as this issue has been demanded by most of the recent grid codes. In Figure 8.22, the DC-link voltage of the wind generator reached almost zero value during the transient state, and it was able to recover. Similarly, the terminal voltage of the DFIG

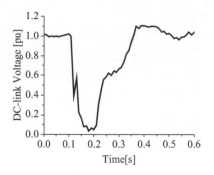

FIGURE 8.22 DC-link voltage of DFIG wind turbine at zero-voltage condition.

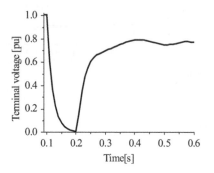

FIGURE 8.23 Terminal voltage of DFIG wind turbine at zero-voltage condition.

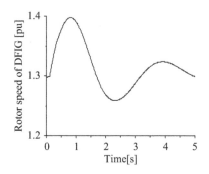

FIGURE 8.24 Rotor speed of DFIG wind turbine at zero-voltage condition.

wind turbine in Figure 8.23 reached zero-voltage, and quickly recovered within the stipulated time set by the grid codes to remain connected to the grid after transient state. The impact of the zero-voltage could also be seen in the response of the wind generator's rotor speed in Figure 8.24. The rotor speed reaches a high oscillation value during the transient state and was able to regain stability within a short time to its steady state.

8.6   CHAPTER CONCLUSION

The use of energy storage elements plays an important role in theoretically resolving transient problems in grid-connected DFIG-based wind turbines. This chapter investigated the effects of a supercapacitor, as an energy storage system, in DFIG transient stability. The supercapacitor was connected at the DC-link voltage, between the rotor side converter and grid side converter of the DFIG wind turbine. The performance of the supercapacitor was investigated by varying its resistance, inductance and capacitance parameters. A simple machine model system of DFIG and fixed speed induction generator tied to an infinite bus was used in the study. The DC-link voltage and grid voltage were used for the switching of the supercapacitor. It was observed that when the resistance, capacitance and inductance parameters of the supercapacitor were too small, the DC-link voltage and grid voltage switching strategies gave poor performances during transient state. However, the performance of the supercapacitor system in the DFIG was improved when the effective values of the parameters during transient state were used. The proposed supercapacitor DFIG scheme was compared to existing solutions in the literature, using the parallel capacitor scheme for the DFIG, considering the same capacitance value. The obtained results show that the use of the existing parallel capacitor scheme in the DFIG grid side power converter was able to enhance the performance of the DC-link voltage and rotor speed of the wind generator, with fast recovery of the variables after transient state, compared to when it is at the rotor side converter of the DFIG. However, the proposed supercapacitor DFIG-based system with optimal parameter ratings gave optimal changes than the conventional DFIG parallel capacitor scheme.

REFERENCES

[1] I. M. Syed, B. Venkatesh, B. Wu, and A. B. Nassif, "Two-layer control scheme for a supercapacitor energy storage system coupled to a doubly fed induction generator," *Electric Power Systems Research*, vol. 86, pp. 76–83, 2012.

[2] Q. Liyan, and W. Qiao, "Constant power control of DFIG wind turbines with supercapacitor energy storage," *IEEE Transactions on Industry Applications*, vol. 47, pp. 359–367, 2011.

[3] V. Krishnamurthy, and C. R. Kumar, "A novel two layer constant power control of 15 DFIG wind turbines with supercapacitor energy storage," *International Journal of Advanced and Innovative Research*, vol. 2, 68–77, 2013.

[4] S. Dongyang, Z. Xiongxin, S. Lizhi, W. Fengjian, and Z. Guangxin, "Study on power fluctuation suppression of DFIG based on super capacitor energy storage," *2017 IEEE Conference on Energy Internet and Energy System Integration (EI2)*, IEEE, November 2017, pp. 1–6.

[5] R. Suryana,"Frequency control of standalone wind turbine with supercapacitor," In *2011 IEEE 33rd International Telecommunications Energy Conference (INTELEC)*, IEEE, October 2011, pp. 1–8.

[6] R. Aghatehrani, R. Kavasseri, and R. C. Thapa, "Power smoothing of the DFIG wind turbine using a small energy storage device," *IEEE PES General Meeting*, IEEE, July 2010, pp. 1–6.

[7] N. Mendis, K. M. Muttaqi, and S. Perera, "Active power management of a supercapacitor-battery hybrid energy storage system for standalone operation of DFIG based wind turbines," *IEEE Industry Applications Society Annual Meeting*, Las Vegas, USA, 7–11 October 2012.

[8] E. Naswali, C. Alexander, H. Y. Han, D. Naviaux, A. Bistrika, A. V. Jouanne, A. Yokochi, and K. A. T. Brekken, "Supercapacitor energy storage for wind energy integration," *IEEE Energy Conversion Congress and Exposition*, Phoenix, Arizona, 17–22 September 2011.

[9] S. Huang, Q. Wu, Y. Guo, and F. Rong, "Optimal active power control based on MPC for DFIG-based wind farm equipped with distributed energy storage systems," *International Journal of Electrical Power & Energy Systems*, vol. 113, pp. 154–163, 2019.

[10] T. Wei, S. Wang, and Z. Qi, "Design of supercapacitor based ride through system for wind turbine pitch systems," *2007 International Conference on Electrical Machines and Systems (ICEMS)*, IEEE, October 2007, pp. 294–297.

[11] K. E. Okedu, Improved Performance of DFIG Wind Turbine during Transient State Considering Supercapacitor Control Strategy, *Electrica Journal*, vol. 22, no. 2, pp. 198–210, 2022. DOI: 10.54614/electrica.2022.21006.

[12] A. B. Cultura, and Z. M. Salameh, "Modeling, evaluation and simulation of a supercapacitor module for energy storage application," *International Conference on Computer Information Systems and Industrial Applications*, Atlantis Press, June 2015.

[13] M. K. Döşoğlu, and A. B. Arsoy, "Transient modeling and analysis of a DFIG based wind farm with supercapacitor energy storage," *International Journal of Electrical Power & Energy Systems*, vol. 78, pp. 414–421, 2016.

[14] M. K. Döşoğlu, "Nonlinear dynamic modeling for fault ride-through capability of DFIG-based wind farm," *Nonlinear Dynamics*, vol. 89, no. 4, pp. 2683–2694, 2017.

[15] R. Isermann, and M. Münchhof, *Identification of Dynamic Systems: An Introduction with Applications*, Berlin, Germany: Springer Science & Business Media, 2010.

[16] Y. Wang, C. Liu, R. Pan, and Z. Chen, "Modeling and state-of-charge prediction of lithium-ion battery and ultracapacitor hybrids with a co-estimator," *Energy*, vol. 121, pp. 739–750, 2017.

[17] M. Ceraolo, G. Lutzemberger, and D. Poli, "State-of-charge evaluation of supercapacitors," *Journal of Energy Storage*, vol. 11, pp. 211–218, 2017.

[18] M. Henry, B. Andrés, F. Cristina, Z. Pablo, and L. Antonio, "A general parameter identification procedure used for the comparative study of supercapacitors models," *Energies*, vol. 12, no. 1776, pp. 1–20, 2019.

[19] *PSCAD/EMTDC Manual*, Manitoba HVDC lab, 2016.

[20] K. E. Okedu, "Enhancing the performance of DFIG variable speed wind turbine using parallel integrated capacitor and modified modulated braking resistor," *IET Generation Transmission & Distribution*, vol.13, no. 15, pp. 3378–3387, 2019.

[21] K. E. Okedu, "Enhancing DFIG wind turbine during three-phase fault using parallel interleaved converters and dynamic resistor," *IET Renewable Power Generation*, vol. 10, no. 6, pp. 1211–1219, 2016.

9 PMSG Wind Turbine with Series and Bridge Fault Current Limiters

9.1 CHAPTER INTRODUCTION

The use of Fault Current Limiters (FCLs) hardware-based solutions in wind turbines has shown good results in fulfilling the Low Voltage Ride Through (LVRT) requirements of grid codes [1]. The technologies of FCLs are basically Superconducting Fault Current Limiter (SFCL) type and Non-superconducting Fault Current Limiter (NSFCL) type. There is no power loss in SFCLs, with good speed control, however, their configurations are complex [2–4]. The technology of the NSFCLs can solve the drawbacks of SFCLs and also provide better LVRT capability [5, 6]. The use of Insulated Gate Bipolar Transistors (IGBTs) gave way for the NSFCLs more than the others. The Series Dynamic Braking Resistors (SDBRs) [7, 8] and Bridge Fault Current Limiters (BFCLs), with resistive, inductive and capacitive elements [9] and parallel resonance type FCL (PRFCL) [10] show considerable improvement of the performance of variable speed wind turbines.

This chapter investigates the performance of PMSG-based wind turbines, considering the SDBR and the BFCL topologies during LVRT. The mathematical dynamics of the power converter for the Grid Side Converter (GSC) of the PMSG were modeled for the SDBR and the BFCL in the PMSG wind turbine for steady and transient states. The robustness of the controllers of the wind generator was tested, using the threshold value of the grid voltage as the switching signal during LVRT for a balanced three-phase-to-ground fault in Power System Computer Aided Design and Electromagnetic Transient Including DC (PSCAD/EMTDC) environment [11].

9.2 OVERVIEW OF FAULT CURRENT LIMITER TOPOLOGIES

There are several types of FCLs used in the LVRT of wind turbines [12, 13]. These FCLs have various benefits and drawbacks, thus, their performance would be different based on the requirements [14]. The FCLs control strategies could be broadly classified into SFCLs, NSFCLs and Magnetic Fault Current Limiters (MFCLs), as shown in Figure 9.1 [15]. Because of their strong ability of mitigating fault currents during transient [16] and no power loss being incurred during their operation, FCLs are currently in use in variable speed wind farms [17].

The SFCL could work in hybrid with Superconducting Magnetic Energy Storage (SMES) for effective energy management during and after transient states [18–20]. However, the SFCLs have complex structures, not cheap to implement, and liquid cryogenic system is required for changing non-superconducting mode to

DOI: 10.1201/9781003350910-9

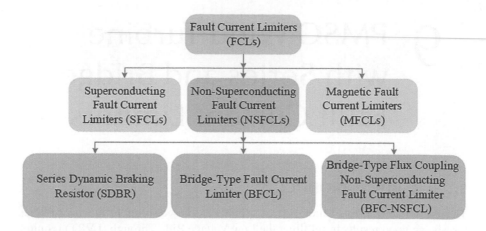

FIGURE 9.1   Classes of fault current limiters.

superconducting mode. The MFCLs configuration has basically a core that is laminated, with a permanent magnet connecting the poles [21]. The use of permanent magnets would help in limiting the high fault currents because of the large square hysteresis loop they possess. The main shortcoming of the MFCLs is that it becomes weak over time, consequently, reducing its efficiency [22, 23].

The NSFCLs are cheaper compared to the SFCLs considering the same level of dynamic stability [24]. The employed NSFCLs [25, 26] have power losses because of their static components. However, SFCLs with flux coupling can reduce the power loss usually found in FCLs. But the high cost incurred would discourage their implementation. In this chapter, the PMSG wind turbine would be investigated during transient state using the SDBR and the BFCL topologies of the NSFCLs.

9.3   MODELING OF THE PMSG WIND TURBINE WITH BACK-TO-BACK CONVERTERS

Figure 9.2 shows the complete system for a grid-connected PMSG wind turbine having full back-to-back converter. The PMSG wind turbine system shown in Figure 9.2 has no gearbox, thus, there is a direct coupling between the wind turbine and the wind generator. The output of the generator is connected to the power grid by a DC-link power converter, MSC and GSC.

For modeling the system dynamics of the PMSG wind turbine, the following differential equations apply. The mechanical power extracted $\left(P_{\mathrm{wt}}\right)$ and the torque $\left(T_{\mathrm{wt}}\right)$ of the PMSG wind turbine are expressed as [27]:

$$P_{\mathrm{wt}} = \frac{1}{2}\rho A C_{\mathrm{P}}\left(\lambda,\beta\right)V^3 \tag{9.1}$$

$$T_{\mathrm{wt}} = \frac{1}{2}\frac{\rho A C_{\mathrm{P}}\left(\lambda,\beta\right)V^3}{\omega_{\mathrm{wt}}} \tag{9.2}$$

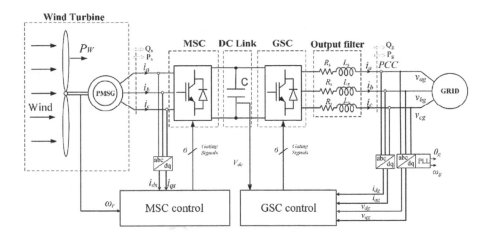

FIGURE 9.2   PMSG wind turbine with back-to-back power converters.

In Equations (9.1) and (9.2), A is the swept area of the blade, ρ is the density of the air, C_p is the coefficient of the rotor, V is the wind speed and ω_{wt} is the rotational speed of the wind turbine. The relationship between the tip speed ratio λ and the blade pitch angle β is given as:

$$\lambda = \frac{\omega_{wt}R}{V} \tag{9.3}$$

The two mass drive train models of the wind turbine mechanical system can be expressed as:

$$T_{gen} = D_m\left(\omega_{wt} - \omega_{gen}\right) + k_{sh}D_{DT} \tag{9.4}$$

$$\omega_{wt} = \frac{1}{2H_{\omega t}}\left(T_{\omega t} - T_{gen}\right) \tag{9.5}$$

where the displacement D_{DT} is given by:

$$D_{DT} = \theta_{\omega t} - \theta_{gen} \tag{9.6}$$

From Equations (9.4) to (9.6), T_{gen} is the mechanical torque of the generator, D_m is the mutual damping coefficient, k_{sh} is the shaft stiffness, $H_{\omega t}$ is the per unit inertia constant, ω_{gen} is the generator's mechanical rotational speed, while $\theta_{\omega t}$ and θ_{gen} are the rotational angles of the wind turbine and the generator.

The voltage equations of the PMSG wind turbine when it operates as a grid-connected generator are given as [28]:

$$v_{ds} = -R_s i_{ds} - L_{ds}\frac{di_{ds}}{dt} + \omega_e L_{qs} i_{qs} \tag{9.7}$$

$$v_{qs} = -R_s i_{qs} - L_{qs}\frac{di_{qs}}{dt} - \omega_e L_{ds} i_{ds} + \omega_e \psi_f \tag{9.8}$$

From Equations (9.7) and (9.8), v_{ds} and v_{qs} are the dq axes stator voltages, i_{ds} and i_{qs} are the dq axes stator currents, L_{ds} and L_{qs} are the dq axes inductances, R_s is the stator resistance, ψ_f is the rotor magnetic flux and ω_e is the electrical angular speed given by:

$$\omega_e = p\omega_{gen} \tag{9.9}$$

where p is the number of pair poles. The electromagnetic torque T_e is expressed as:

$$T_e = p(\psi_f i_{qs} + (L_{ds} - L_{qs})i_{ds}i_{qs} \tag{9.10}$$

The mechanical equation for the PMSG is:

$$T_{gen} - T_e = J\frac{d\omega_{gen}}{dt} \tag{9.11}$$

In Equation (9.11), J is the total mechanical system moment of inertia. The voltage of the DC-link is affected by the extracted power flow of the generator and the injected power flow of the grid. Neglecting the power loss of the generator and the converter loss, the dynamics of the DC-link voltage V_{dc} is given as [29]:

$$CV_{dc}\frac{dV_c}{dt} = P_{gen} - P_{grid} = T_e\omega_{gen} - P_{grid} \tag{9.12}$$

The PMSG wind turbine characteristics are shown in Figure 9.3, for the turbine output power with different speeds, where the reference power P_{ref} is based on the rated power. The maximum power output of 1.0 pu was obtained at 12 m/s and 1.0 pu rotational speed.

9.4   THE POWER CONVERTER CONTROL OF THE PMSG WIND TURBINE

Figure 9.4 shows the power converter control for the MSC and the GSC of the PMSG wind turbine. In normal conditions, the MSC regulates the active power control to extract maximum wind power. The MSC of the PMSG wind turbine is shown in Figure 9.4, where the rotor speed is regulated based on its reference value obtained from the MPPT characteristics in Figure 9.3. In Figure 9.4, the q-axis current (i_{qsref}) is generated from the Proportional Integral (PI) controller and the component in the d-axis (i_{dsref}) that is set to zero value. This control topology would help to achieve the required reference values, so that maximum power could be extracted and delivered to the DC-link of the PMSG wind turbine.

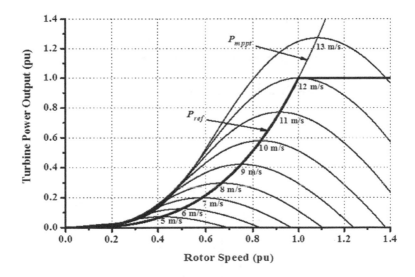

FIGURE 9.3 The maximum power characteristics of the PMSG wind turbine.

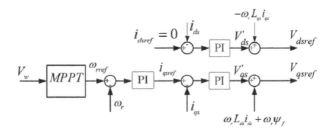

FIGURE 9.4 The PMSG wind turbine machine side converter.

On the other hand, the control strategy for the GSC is shown in Figure 9.5, where after obtaining the MPPT from the MSC, the reference d-axis grid current $\left(i_{\text{dgref}}\right)$ is generated by the DC-link voltage controllers, in order to adjust the terminal voltage to its steady state value. The active and reactive power which are the injected grid power components are controlled independently. The control of the dq components is done through the PI controllers' loop that is cascaded in nature.

9.5 THE PMSG MODEL SYSTEM WITH SDBR AND BFCL

The model system of this study is shown in Figure 9.6. A system base of 5.0 MVA and short circuit of 16.67 MVA were used in the study and the PMSG wind turbine is connected to an infinite bus. The model system parameters are shown in Table 9.1 [30, 31]. On the double circuit of the model system in Figure 9.6, a severe three-phase-to-ground fault occurred. The SDBR and BFCL are connected to the GSC of the PMSG wind turbine. The connection of either the SDBR or BFCL to the PMSG

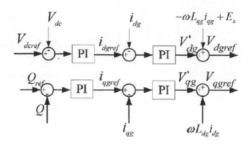

FIGURE 9.5 The PMSG wind turbine grid side converter.

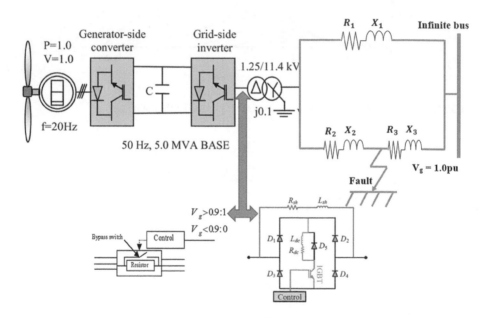

FIGURE 9.6 Model system of PMSG wind turbine with SDBR and BFCL.

TABLE 9.1

Parameters of the Model System

Rated power	5.0 MW	Rated voltage	1.0 kV
Rated voltage	1.0 kV	Field flux	1.4 pu
Frequency	20.0 Hz	Blade radius	40.0 m
Number of poles	150.0	Rated wind speed	12.0 m/s
Machine Inertia	3.0	R_1	0.87120 Ω
Stator resistance	0.01 pu	R_2	0.04356 Ω
d-axis reactance	1.0 pu	R_3	0.82764 Ω
q-axis reactance	0.7 pu	X_1	5.2157 Ω
X_2	0.2608 Ω	X_3	4.9549 Ω

TABLE 9.2

Parameters of the Fault Current Limiters

SDBR		BFCL			
Series resistance (R_s) 0.1 pu		R_{sh}	L_{sh}	R_{dc}	L_{dc}
		20 Ω	250 mH	0.003 Ω	1 mH

wind turbine GSC would enhance its performance during grid fault. The parameters of the SDBR and BFCL used in this study are given in Table 9.2. The switching of the SDBR and BFCL is based on the grid voltage, during steady state (above 0.9 pu) or grid disturbance (below 0.9 pu). The analyses of SDBR and BFCL mathematical dynamics in the PMSG wind turbine are as follows.

The connection of the SDBR and BFCL in the PMSG-based wind turbine shown in Figure 9.6 is based on current control topology and not voltage [32, 33]. The resistor is usually bypassed during nominal operation when the switch is conducting, based on the threshold value (0.9 pu) of the grid voltage used for switching purposes. However, during fault scenario, the switch is off. The SDBR and BFCL would limit the high rotor inrush current while operating, thus excessive active power would be achieved [34, 35]. The MSC and GSC power converters would be effectively balanced, mitigating the current in the stator and the DC-link capacitor charging, as a result of the accumulation of these effects on the PMSG wind turbine.

The control structures of the SDBR and the BFCL are shown in Figure 9.6 model system. The BFCL is made up of two parts: a typical bridge circuit with four diodes (D_1–D_4) and a shunt path made up of inductor (L_{sh}) and resistor (R_{sh}) in series. There is an IGBT switch connected in series with an inductor (L_{dc}) and (R_{dc}) act as an intrinsic resistance of (L_{dc}) with a very small magnitude that is negligible. In the BFCL, the (L_{dc}) inductor is a DC reactor due to the fact that current flows in one direction only during its positive and negative half cycle of the alternating current. There is a freewheeling diode D_5 that is connected to the DC reactor to protect the system from inductive kick during transient state [36]. The principle of the BFCL during operation is such that at steady state, the current flows through the $D1$-Ldc-Rdc-IGBT-D_4 path for the positive half cycle and through the $D3$-IGBT-Rdc-Ldc-D_2 path for the negative half cycle. The shunt path of the BFCL has a very high impedance, thus, the line current and some negligible leakage currents are on the bridge switch [37, 38].

The topology of the voltage source converter in wind energy conversion results in high DC-link voltage quality, capability of bidirectional flow of power, with power factor that is unity and few distortions of the current. Because the technology of the PMSG is fully decoupled from the power grid as a result of its back-to-back power converter, inserting the SDBR and BFCL with the grid-connected inverter in Figure 9.7 would help understand the dynamics of the SDBR and BFCL in the PMSG wind turbine GSC.

From Figure 9.7, the GSC of the PMSG is connected to the R and L of the power grid, having currents $i = a,b,c$. If C_{abc} represents the three switching states for the IGBTs, then the C_{abc} can be replaced by β_{abc} modulation signals. Considering Park's transformation, the PMSG converter could be modeled for a balanced three-phase as [37]:

$$e_d = -\omega L i_q + L\frac{di_d}{dt} + \left(R + R_{SDBR} \text{ or } Z_{BFCL}\right)i_d + 0.5U_{dc}\beta_d \tag{9.13}$$

$$e_q = -\omega L i_d + L\frac{di_q}{dt} + \left(R + R_{SDBR} \text{ or } Z_{BFCL}\right)i_q + 0.5U_{dc}\beta_q \tag{9.14}$$

$$C\frac{dU_{dc}}{dt} = 0.75\left(i_d\beta_d + i_q\beta_q\right) - \frac{U_{dc}}{R_L} \tag{9.15}$$

$$r = \sqrt{\beta_d^2 + \beta_q^2} \tag{9.16}$$

From Equations (9.13) to (9.16), i_d, i_q are dq current input of the rectifier's axes; e_d, e_q are dq voltage of the grid voltage axes components; ω is the angular frequency voltage; β_d, β_q are the rectifier's d and q axes components, while r is the modulation signal vector norm; U_{dc}, is the DC-link voltage; R_{SDBR}, is the effective SDBR resistance; Z_{BFCL} is the effective BFCL reactance and ω is the angular frequency. The Park's principles for three-phase transformation, for phase-A grid voltage with the dq reference, are:

$$e_d = E_m \tag{9.17}$$

$$e_q = 0 \tag{9.18}$$

E_m is the voltage amplitude, e_d and e_q, the d and q source voltages. The active $\left(P_s\right)$ and reactive (Q_s) rectifier's powers are:

$$P_s = \frac{3}{2}E_m i_d \tag{9.19}$$

$$Q_s = -\frac{3}{2}E_m i_q \tag{9.20}$$

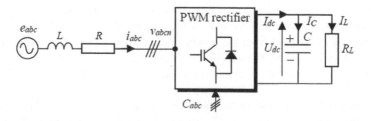

FIGURE 9.7   Topologies of SDBR and BFCL in PMSG.

In order to obtain power factor of 1, i_{q_ref} should be 0. Therefore, for the current regulation to be ideal, $i_q = i_{qref} = 0$. Considering $i_q = 0$ and $e_q = 0$, voltage source converter for power factor 1 is:

$$E_m = L\frac{di_d}{dt} + (R + R_{SDBR} \text{ or } Z_{BFCL})i_d + 0.5U_{dc}\beta_d \tag{9.21}$$

$$\beta_q = -\frac{2\omega L}{U_{dc}}i_q \tag{9.22}$$

$$C\frac{dU_{dc}}{dt} = \frac{3}{4}i_d\beta_d - \frac{U_{dc}}{R_L} \tag{9.23}$$

For power factor of 1, the voltage source converter, β_q should vary with i_q current. Thus, the capacitor charge is manipulated by β_d, via the i_d current of the input based on Equations (9.21) and (9.23). When the SDBR and BFCL are inserted, Equations (9.21)–(9.23) would be zero, making:

$$E_m = (R + R_{SDBR} \text{ or } Z_{BFCL})i_d + 0.5U_{dc}\beta_d \tag{9.24}$$

$$\beta_q = -\frac{2\omega L}{U_{dc}}i_q \tag{9.25}$$

$$i_d = \frac{4U_{dc}}{3\beta_d R_L} \tag{9.26}$$

For a load β_d and voltage $\beta_d = \beta_{d2}$, is

$$6E_m R_L \beta_d - 8(R + R_{SDBR} \text{ or } Z_{BFCL})U_{dc} - 3R_L\beta_d^2 U_{dc} = 0, \text{ for } \beta_d \neq 0 \tag{9.27}$$

Giving two solutions:

$$\beta_{d1} = \frac{E_m}{U_{dc}} - \sqrt{\left(\frac{E_m}{U_{dc}}\right)^2 - \frac{8(R + R_{SDBR} \text{ or } Z_{BFCL})}{3R_L}} \tag{9.28}$$

$$\beta_{d2} = \frac{E_m}{U_{dc}} + \sqrt{\left(\frac{E_m}{U_{dc}}\right)^2 - \frac{8(R + R_{SDBR} \text{ or } Z_{BFCL})}{3R_L}} \tag{9.29}$$

Solution of β_d in Equation (9.28) is not feasible because it has very low values. However, the solution of Equation (9.29) is acceptable, making $\beta_d = \beta_{d2}$, and β_d would exist if:

$$\left(\frac{E_m}{U_{dc}}\right)^2 - \frac{8(R + R_{SDBR} \text{ or } Z_{BFCL})}{3R_L} \geq 0 \tag{9.30}$$

$$\text{But} \quad P_{dc} \leq P_{dc\_max} \tag{9.31}$$

where P_{dc_max} is the PMSG converter maximum, and from the power conservation principle, P_{dc_max} could be expressed as:

$$P_{dc} = \frac{3}{2}E_m i_d - \frac{3}{2}\left(R + R_{SDBR} \text{ or } Z_{BFCL}\right)i_d^2 \tag{9.32}$$

If $dP_{dc} / di_d = 0$, the maximum power transfer to the DC would be:

$$\frac{dP_{dc}}{di_d} = \frac{3}{2}E_m - 3Ri_d = 0 \rightarrow i_d = i_{d\_max} = \frac{E_m}{2R} \tag{9.33}$$

Substituting Equation (9.33) into (9.32):

$$P_{dc\_max} = \frac{3E_m^2}{8\left(R + R_{SDBR} \text{ or } Z_{BFCL}\right)} \tag{9.34}$$

And the voltage source converter operation is possible when

$$P_{dc} \leq P_{dc\_max} \rightarrow \left(\frac{E_m}{U_{dc}}\right)^2 - \frac{8\left(R + R_{SDBR} \text{ or } Z_{BFCL}\right)}{3R_L} \geq 0 \tag{9.35}$$

The grid input maximal power P_{s_max} can be obtained by putting Equation (9.34) into Equation (9.20) for Q_s. Thus:

$$P_{s\_max} = \frac{3E_m^2}{4\left(R + R_{SDBR} \text{ or } Z_{BFCL}\right)} \tag{9.36}$$

Therefore, during transient state, the maximum power transfer in the GSC of the PMSG and the total current would be reduced. The oscillations that usually occur would also be less because of the control strategies of the SDBR and the BFCL.

9.6 EVALUATION OF THE SYSTEM PERFORMANCE

The model system of the PMSG wind turbine with the SDBR and BFCL was carried out in PSCAD/EMTDC environment. The model system was subjected to a severe symmetrical three-phase fault of 100 ms happening at 10.1 s, with the circuit breakers operation sequence opening and reclosing at 10.2 s and 11 s, respectively. The model system was evaluated considering three scenarios. In the first scenario, no control strategy of FCL was employed in the PMSG wind turbine. In the second scenario, the SDBR FCL control strategy was used in the PMSG wind turbine, while the BFCL was implemented in the third scenario. The PMSG wind turbine was operating at its rated speed during the grid fault. Some of the PMSG wind turbine variables are shown in Figures 9.8–9.12, for the three scenarios considered in this study.

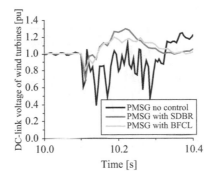

FIGURE 9.8   DC-link voltage of the PMSG wind turbine.

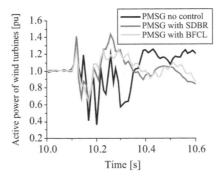

FIGURE 9.9   Active power of the PMSG wind turbine.

Figures 9.8 and 9.9 show the DC-link voltage and active power of the PMSG wind turbine for no control strategy, using FCL with SDBR, and with BFCL control strategies. It is obvious from Figures 9.8 and 9.9 that connecting SDBR and BFCL on the PMSG wind turbine GSC would improve the DC-link voltage and active power variables during grid fault because of the decoupling of the PMSG wind turbine by the grid based on its back-to-back power converter. Using the metrics of the overshoot, undershoot and settling time of the variables of the wind generator, scenarios where the SDBR and BFCL were implemented gave better responses than no control scenario. One of the technical reasons for this is based on the fact that connecting the SDBR and BFCL on GSC of the PMSG divides the expected high voltage in the stator circuitry of the PMSG wind turbine, since it is a series connection topology, based on the mathematical dynamics of their connection earlier explained. The performance of the BFCL is better than SDBR because of the additional energy buffer from the inductive circuit of the BFCL.

Figure 9.10 shows that the reactive power was better controlled and dissipated during transient state using the BFCL than the SDBR and no control scenarios. This is because the inductive circuit would act as an energy buffer to the PMSG wind turbine, thereby controlling its reactive power. Consequently, the terminal voltage of the PMSG wind turbine would be improved, as shown in Figure 9.11, since reactive

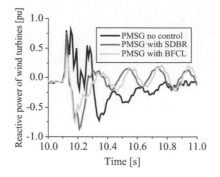

FIGURE 9.10   Reactive power of the PMSG wind turbine.

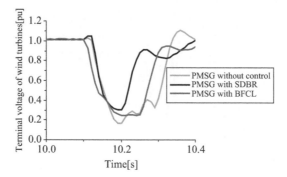

FIGURE 9.11   Terminal voltage of the PMSG wind turbine.

power is directly proportional to voltage. The voltage variable of the PMSG wind turbine settled faster using the BFCL than the SDBR and no control scenarios. Though the SDBR gave low voltage dip and the same overshoot as the BFCL, however, its settling time is more than that of the BFCL control strategy.

In Figure 9.12, the performance of the rotor speed of the PMSG wind turbine is better with the use of SDBR and BFCL, than the scenario without control. The rotor

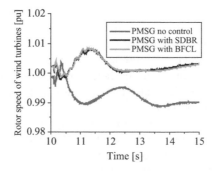

FIGURE 9.12   Rotor speed of the PMSG wind turbine.

speed responses for the SDBR and BFCL scenarios are the same because their control techniques in the PMSG wind turbine have the capability to enhance the mechanical output and at the same time limit the speed during grid disturbances. These effects would lead to fewer oscillations and improved settling time. Based on the presented results, employing the SDBR and BFCL at the GSC of the PMSG wind turbine gives room for no induced overvoltage, no loss of power converter control, mitigation of high current flow and no excessive charging current in the capacitor of the DC-link voltage of the wind generator.

9.7   CHAPTER CONCLUSION

In this chapter, the performance of the PMSG wind turbine was investigated using SDBR and BFCL. The grid voltage set value during grid fault was used as the switching signal for the IGBTs of both FCLs, for fair comparison. A severe symmetrical three-phase-to-ground fault at the terminals of the PMSG was applied to test the robustness of both protection schemes. Three scenarios were investigated for the PMSG wind turbine: no control strategy, with SDBR at the GSC and with BFCL at the GSC. When no control was implemented, there were substantial effects on the PMSG during transient state. Apparently, using the SDBR and BFCL would enhance the PMSG wind turbine performance. The performance of the BFCL was superior to that of the SDBR during transient state. Therefore, the BFCL would help solve fault ride through capability of PMSG wind farms even though its structure is complex, more than the simple SDBR topology.

REFERENCES

[1] M. Islam, M. Huda, J. Hasan, M. A. H. Sadi, A. AbuHussein, T. K. Roy, and M. Mahmud, "Fault ride through capability improvement of DFIG based wind farm using nonlinear controller based bridge-type flux coupling non-superconducting fault current limiter," *Energies*, vol. 13, no. 7, p. 1696, 2020.

[2] M. Firouzi, "Low-voltage ride-through (LVRT) capability enhancement of DFIG-based wind farm by using bridge-type superconducting fault current limiter (BTSFCL)," *Journal of Power Technologies*, vol. 99, no. 4, pp. 245–253, 2020.

[3] M. R. Islam, J. Hasan, M. M. Hasan, M. N. Huda, M. A. H. Sadi, and A. AbuHussein, "Performance improvement of DFIG-based wind farms using narma-l2 controlled bridge-type flux coupling non-superconducting fault current limiter," *IET Generation, Transmission & Distribution*, vol. 14, no. 26, pp. 6580–6593, 2021.

[4] J. Hasan, M. R. Islam, M. R. Islam, A. Z. Kouzani, and M. P. Mahmud, "A capacitive bridge-type superconducting fault current limiter to improve the transient performance of DFIG/PV/SG-based hybrid power system," *IEEE Transactions on Applied Superconductivity*, 2021, doi: 10.1109/TASC.2021.3094422.

[5] G. Rashid, and M. H. Ali, "A modified bridge-type fault current limiter for fault ride-through capacity enhancement of fixed speed wind generator," *IEEE Transactions on Energy Conversion*, vol. 29, no. 2, pp. 527–534, 2014.

[6] M. M. Moghimian, M. Radmehr, and M. Firouzi, "Series resonance fault current limiter (SRFCL) with mov for LVRT enhancement in DFIG-based wind farms," *Electric Power Components and Systems*, vol. 47, no. 19–20, pp. 1814–1825, 2019.

[7] J. Yang, J. E. Fletcher, and J. O'Reilly, "A series-dynamic-resistor-based converter protection scheme for doubly-fed induction generator during various fault conditions," *IEEE Transactions on Energy Conversion*, vol. 25, no. 2, pp. 422–432, 2010.

[8] Z. Din, J. Zhang, Z. Xu, Y. Zhang, and J. Zhao, "Low voltage and high voltage ride-through technologies for doubly fed induction generator system: Comprehensive review and future trends," *IET Renewable Power Generation*, vol. 15, no. 3, pp. 614–630, 2021.

[9] M. Firouzi, and G. Gharehpetian, "Improving fault ride-through capability of fixed-speed wind turbine by using bridge-type fault current limiter," *IEEE Transactions on Energy Conversion*, vol. 28, no. 2, pp. 361–369, 2013.

[10] S. B. Naderi, M. Jafari, and M. T. Hagh, "Parallel-resonance-type fault current limiter," *IEEE Transactions on Industrial Electronics* vol. 60, no. 7, pp. 2538–2546, 2012.

[11] PSCAD/EMTDC Manual, Manitoba HVDC lab, 2016.

[12] H. Radmanesh, S. H. Fathi, G. Gharehpetian, and A. Heidary, "Bridge-type solid-state fault current limiter based on AC/DC reactor," *IEEE Transactions on Power Delivery*, vol. 31, pp. 200–209, 2015.

[13] H. Shen, F. Mei, J. Zheng, H. Sha, and C. She, "Three-phase saturated-core fault current limiter," *Energies*, vol. 11, pp. 3471, 2018.

[14] M. Alam, M. Abido, and I. El-Amin, "Fault current limiters in power systems: A comprehensive review," *Energies*, vol. 11, pp. 1025, 2018.

[15] M. Islam, M. Huda, J. Hasan, M. A. H. Sadi, A. AbuHussein, T. K. Roy, and M. Mahmud, "Fault ride through capability improvement of DFIG based wind farm using nonlinear controller based bridge-type flux coupling non-superconducting fault current limiter," *Energies*, vol. 13, no. 7, p. 1696, 2020.

[16] Q. Yang, S. Le Blond, F. Liang, W. Yuan, M. Zhang, and J. Li, "Design and application of superconducting fault current limiter in a multiterminal HVDC system," *IEEE Transactions on Applied Superconductivity*, vol. 27, pp. 1–5, 2017.

[17] Z. C. Zou, X. Y. Xiao, Y. F. Liu, Y. Zhang, and Y. H. Wang, "Integrated protection of DFIG-based wind turbine with a resistive-type SFCL under symmetrical and asymmetrical faults," *IEEE Transactions on Applied Superconductivity*, vol. 26, pp. 1–5, 2016.

[18] I. Ngamroo, and T. Karaipoom, "Improving low-voltage ride-through performance and alleviating power fluctuation of DFIG wind turbine in DC microgrid by optimal SMES with fault current limiting function," *IEEE Transactions on Applied Superconductivity* vol. 24, pp. 1–5, 2014.

[19] H. J. Lee, S. H. Lim, and J. C. Kim, "Application of a superconducting fault current limiter to enhance the low-voltage ride-through capability of wind turbine generators," *Energies*, vol. 12, pp. 1478, 2019.

[20] Z. C. Zou, X. Y. Chen, C. S. Li, X. Y. Xiao, and Y. Zhang, "Conceptual design and evaluation of a resistive-type SFCL for efficient fault ride through in a DFIG," *IEEE Transactions on Applied Superconductivity*, vol. 26, pp. 1–9, 2015.

[21] J. Rasolonjanahary, J. Sturgess, E. Chong, A. Baker, and C. Sasse, Design and construction of a magnetic fault current limiter. *Proceedings of the 3rd IET International Conference on Power Electronics, Machines and Drives-PEMD*, Dublin, Ireland, 4–6 April 2006; London, UK: IET, pp. 681–685, 2006.

[22] L. Qu, R. Zeng, Z. Yu, and G. Li, "Design and test of a magnetic saturation-type fault current limiter," *Journal of Engineering*, vol. 2019, no. 16, pp. 2974–2979, 2019.

[23] J. Yuan, Y. Zhong, Y. Lei, C. Tian, W. Guan, Y. Gao, K. Muramatsu, and B. Chen, "A novel hybrid saturated core fault current limiter topology considering permanent magnet stability and performance," *IEEE Transactions on Magnetics*, vol. 53, pp. 1–4, 2017.

[24] H. T. Tseng, W. Z. Jiang, and J. S. Lai, "A modified bridge switch-type flux-coupling nonsuperconducting fault current limiter for suppression of fault transients," *IEEE Transactions on Power Delivery*, vol. 33, pp. 2624–2633, 2018.

[25] G. Rashid, and M. H. Ali, "Transient stability enhancement of doubly fed induction machine-based wind generator by bridge-type fault current limiter," *IEEE Transactions on Energy Conversion*, vol. 30, pp. 939–947, 2015.

[26] M. A. H. Sadi, and M. H. Ali, "A fuzzy logic controlled bridge type fault current limiter for transient stability augmentation of multi-machine power system," *IEEE Transactions on Power Systems*, vol. 31, pp. 602–611, 2015.

[27] I. Sami, S. U. N. Uiiah, and J.-S. Ro, "Sensorless fractional order composite sliding mode control design for wind generation system," *ISA Transactions*, vol. 111, pp. 275–289, 2021.

[28] S. Li, T. Haskew, and L. Xu, "Conventional and novel control designs for direct driven PMSG wind turbines," *Electric Power Systems Research*, vol. 80, pp. 328–338, 2010.

[29] H. M. Yassin, H. H. Hanafy, and M. M. Hallouda, "Low voltage ride-through technique for PMSG wind turbine systems using interval type-2 fuzzy logic control," *The 16th EEE International Conference on Industrial Technology (ICIT)*, Seville, Spain, 2015.

[30] K. E. Okedu, and Hind Barghash, "Enhancing the transient state performance of permanent magnet synchronous generator based variable speed wind turbines using power converters excitation parameters," *Frontiers in Energy Research-Smart Grids*, vol. 9, pp. 109–120, Article 655051, 2021, doi: 10.3389/fenrg.2021.655051

[31] K. E. Okedu, and S. M. Muyeen, "Enhanced performance of PMSG Wind Turbines during grid disturbance at different network strengths considering fault current limiter," *International Transactions on Electrical Energy Systems, Wiley*, vol. e12985, no. 6, pp. 1–21, 2021, doi: 10.1002/2050-7038.12985.

[32] K. E. Okedu, S. M. Muyeen, R. Takahashi, and J. Tamura, "Wind farms fault ride through using DFIG with new protection scheme," *IEEE Transactions on Sustainable Energy*, vol. 3 no. 2, pp. 242–254, 2012.

[33] K. E. Okedu, "Effect of ECS low pass filter timing on grid frequency dynamics of a power network considering wind energy penetration," *IET Renewable Power Generation*, vol. 11, no. 9, pp. 1194–1199, 2017.

[34] K. E. Okedu, "Determination of the most effective switching signal and position of braking resistor in DFIG wind turbine under transient conditions," *Electrical Engineering*, vol. 102, no. 11, pp.471–480, 2020.

[35] K. E. Okedu, S. M. Muyeen, R. Takahashi, and J. Tamura, "Wind farm stabilization by using DFIG with current controlled voltage source converters taking grid codes into consideration," *IEEJ Transactions on Power and Energy*, vol. 132, no. 3. pp. 251–259, 2012.

[36] K. E. Okedu, S. M. Muyeen, R. Takahashi, and J. Tamura, "Improvement of fault ride through capability of wind farm using DFIG considering SDBR," *14th European Conference of Power Electronics EPE*, Birmingham, United Kingdom, August 2011, pp. 1–10.

[37] M. Islam, M. Huda, J. Hasan, M. A. H. Sadi, A. AbuHussein, T. K. Roy, and M. Mahmud, "Fault ride through capability improvement of DFIG based wind farm using nonlinear controller based bridge-type flux coupling non-superconducting fault current limiter," *Energies* vol. 13, no. 7, p. 1696, 2020.

[38] G. Rashid, and M. H. Ali, "Fault ride through capability improvement of DFIG based wind farm by fuzzy logic controlled parallel resonance fault current limiter," *Electric Power Systems Research*, vol. 146, pp. 1–8, 2017.

10 PMSG with Capacitive Bridge Fault Current Limiters

10.1 CHAPTER INTRODUCTION

Fault Current Limiters (FCLs) are hardware-based solutions in wind turbines and they have proven to be one of the best techniques for fulfilling the Fault Ride Through (FRT) or LVRT requirements, as set by the grid codes [1]. The technology of FCLs comprises two categories: SFCL and Non-superconducting Fault Current Limiter (NSFCL). There is no loss of power during nominal operation in SFCLs, and very high-speed control could be achieved. Though complex configuration may be required for maintenance purposes in this type of FRT solution [2–4]. However, the NSFCLs technology can effectively compensate for the shortcomings of SFCLs and at the same time improve the LVRT capability [5, 6]. The use of semiconductor devices such as Silicon Controlled Rectifier (SCR) and Insulated Gate Bipolar Transistor (IGBT) gave way for the NSFCLs more than others. Consequently, the control strategies of Series Dynamic Braking Resistors (SDBRs) [7, 8], Bridge Fault Current Limiters (BFCLs) with resistive, inductive and capacitive elements [5, 9], series resonance type FCL [6] and Parallel Resonance Type FCL (PRFCL) [10] show improved performance of variable speed DFIG-based wind turbines. Among the BFCLs, the Capacitive Bridge-type Fault Current Limiter (CBFCL) is newly introduced to enhance the traditional BFCL and the FRT of wind turbines [11–13]. One of the main reasons for this could be due to the fact that it provides reasonable reactive power that is required to recover the terminal voltage of the wind generators and the entire system during transient state, compared to the other BFCLs.

This chapter targets the improved performance of PMSG-based wind generators, considering different control topologies of FCL. The FCL considered are the SDBR control strategy, the traditional BFCL and the CBFCL. The mathematical dynamics of the three FCLs in the PMSG wind turbine were presented during steady and transient states of the wind turbine. The same switching strategy based on the grid voltage during fault condition was used for all three FCLs for fair comparison. The robustness of the controllers of the PMSG wind turbine was tested using severe three-phase-to-ground fault in Power System Computer Aided Design and Electromagnetic Transient Including DC (PSCAD/EMTDC) environment.

DOI: 10.1201/9781003350910-10

10.2   THE PMSG MODEL SYSTEM WITH THE DIFFERENT FAULT CURRENT LIMITERS

The model system of this study is shown in Figure 10.1, where the PMSG wind turbine is connected to an infinite bus, with a system base of 5.0 MVA and a short circuit of 16.67 MVA. The parameters of the model system are given in Table 10.1 [14, 15]. A severe balanced three-phase-to-ground fault occurred on the double circuit of the model system. The three FCLs are connected to the GSC of the PMSG wind turbine as shown in the model system. The connection of the FCLs to the PMSG wind turbine would improve its performance during transient state. The effective parameters of the FCLs are given in Table 10.2. The dynamics of the three FCLs in the PMSG wind turbine are given in the subsequent subsections. The switching of the three FCLs is based on the grid voltage, as shown in the model system, during normal state when the grid voltage is above 0.9 pu and during fault conditions when it is less than 0.9 pu.

10.2.1   PMSG WIND TURBINE WITH SDBR

The connection of the SDBR in the PMSG-based wind turbine is shown in Figure 10.2. SDBR control strategy is based on current and not voltage [16, 17]. The resistor is bypassed during nominal operation when the switch is conducting, based on the threshold value of the grid voltage. However, during fault scenario, the switch is off. The switching strategy is based on the grid voltage, as explained earlier in Section 10.2 and shown in Figure 10.1. The SDBR would limit the high rotor inrush current while operating, thus excessive active power would be achieved [18, 19]. Due to these effects, the MSC and GSC power converters would be effectively balanced, reducing the current in the stator and the DC-link capacitor charging.

The GSC of the PMSG wind turbine is connected to the R and L parameters of the grid, with AC currents $i = a, b, c$. If C_{abc} represents the three switching states for the IGBTs, then the C_{abc} converter functions can be substituted by β_{abc} signals of modulation. Considering Park's transformation, the voltage source converter of the PMSG could be modeled for a balanced three-phase as [20]:

$$e_d = -\omega L i_q + L\frac{di_d}{dt} + \left(R + R_{SDBR}\right)i_d + 0.5U_{dc}\beta_d \tag{10.1}$$

$$e_q = -\omega L i_d + L\frac{di_q}{dt} + \left(R + R_{SDBR}\right)i_q + 0.5U_{dc}\beta_q \tag{10.2}$$

$$C\frac{dU_{dc}}{dt} = 0.75\left(i_d\beta_d + i_q\beta_q\right) - \frac{U_{dc}}{R_L} \tag{10.3}$$

$$r = \sqrt{\beta_d^2 + \beta_q^2} \tag{10.4}$$

where i_d, i_q are dq current input of the rectifier's axes, e_d, e_q are dq voltage of the grid voltage axes components, ω is the angular frequency voltage, β_d, β_q are the rectifier's

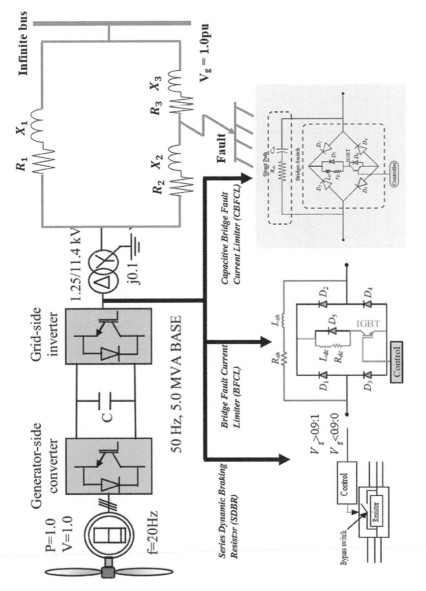

FIGURE 10.1 PMSG wind turbine with different fault current limiters.

TABLE 10.1

Parameters of the Model System

Rated power	5.0 MW	Rated voltage	1.0 kV
Rated voltage	1.0 kV	Field flux	1.4 pu
Frequency	20.0 Hz	Blade radius	40.0 m
Number of poles	150.0	Rated wind speed	12.0 m/s
Machine inertia	3.0	R_1	0.87120 Ω
Stator resistance	0.01 pu	R_2	0.04356 Ω
d-axis reactance	1.0 pu	R_3	0.82764 Ω
q-axis reactance	0.7 pu	X_1	5.2157 Ω
X_2	0.2608 Ω	X_3	4.9549 Ω

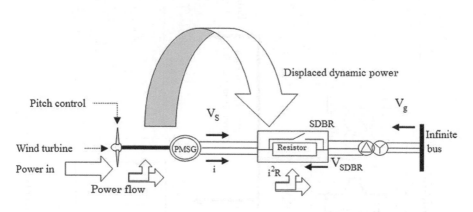

FIGURE 10.2   Dynamics of SDBR at stator side of PMSG wind turbine.

TABLE 10.2

Parameters of the Fault Current Limiters

SDBR	BFCL				CBFCL			
Series	R_{sh}	L_{sh}	R_{dc}	L_{dc}	R_{sh}	L_{sh}	R_{dc}	C_{sh}
resistance	20 Ω	250 mH	0.003 Ω	1 mH	20 Ω	250 mH	0.003 Ω	69 μF
(R_s) 0.1 pu								

d and q axes components, while r is the modulation signal vector norm, U_{dc}, is the DC-link voltage, R_{SDBR}, is the effective SDBR resistance and ω is the angular frequency. Park's principles for three-phase transformation, for phase-A grid voltage with the dq reference, are

$$e_d = E_m \tag{10.5}$$

$$e_q = 0 \tag{10.6}$$

E_m is the voltage amplitude, e_d and e_q, the d and q source voltages. The active (P_s) and reactive (Q_s) rectifier's powers are:

$$P_s = \frac{3}{2} E_m i_d \tag{10.7}$$

$$Q_s = -\frac{3}{2} E_m i_q \tag{10.8}$$

For unity power factor, i_{q_ref} is 0. Therefore, for the current regulation to be ideal, $i_q = i_{qref} = 0$. Considering $i_q = 0$ and $e_q = 0$, voltage source converter for unity power factor is:

$$E_m = L\frac{di_d}{dt} + \left(R + R_{SDBR}\right)i_d + 0.5U_{dc}\beta_d \tag{10.9}$$

$$\beta_q = -\frac{2\omega L}{U_{dc}} i_q \tag{10.10}$$

$$C\frac{dU_{dc}}{dt} = \frac{3}{4}i_d\beta_d - \frac{U_{dc}}{R_L} \tag{10.11}$$

For unity power factor of the voltage source converter, β_q should vary with i_q current. Thus, the capacitor charge is manipulated by β_d, via the i_d current of the input based on Equations (10.9) and (10.11). With the connection of the SDBR, Equations (10.9)–(10.11) would be zero, making:

$$E_m = \left(R + R_{SDBR}\right)i_d + 0.5U_{dc}\beta_d \tag{10.12}$$

$$\beta_q = -\frac{2\omega L}{U_{dc}} i_q \tag{10.13}$$

$$i_d = \frac{4U_{dc}}{3\beta_d R_L} \tag{10.14}$$

For a load R_L and voltage U_{dc}, β_d, is

$$6E_m R_L\beta_d - 8\left(R + R_{SDBR}\right)U_{dc} - 3R_L\beta_d^2 U_{dc} = 0, \text{ for } \beta_d \neq 0 \tag{10.15}$$

Leading to two solutions:

$$\beta_{d1} = \frac{E_m}{U_{dc}} - \sqrt{\left(\frac{E_m}{U_{dc}}\right)^2 - \frac{8\left(R + R_{SDBR}\right)}{3R_L}} \tag{10.16}$$

$$\beta_{d2} = \frac{E_m}{U_{dc}} + \sqrt{\left(\frac{E_m}{U_{dc}}\right)^2 - \frac{8\left(R + R_{SDBR}\right)}{3R_L}} \tag{10.17}$$

Solution of R_L in Equation (10.16) is not feasible because it has very low values. However, the solution of Equation (10.17) is the acceptable, making $\beta_d = \beta_{d2}$, and β_d would exist if:

$$\left(\frac{E_{\mathrm{m}}}{U_{\mathrm{dc}}}\right)^2 - \frac{8\left(R + R_{\mathrm{SDBR}}\right)}{3R_{\mathrm{L}}} \geq 0 \tag{10.18}$$

$$\text{But} \quad P_{dc} \leq P_{dc\_\max} \tag{10.19}$$

where $P_{dc\_\max}$ is the PMSG converter maximum, and from the power conservation principle, $P_{dc\_\max}$ could be expressed as:

$$P_{dc} = \frac{3}{2}E_{\mathrm{m}}i_d - \frac{3}{2}\left(R + R_{\mathrm{SDBR}}\right)i_d^2 \tag{10.20}$$

If $dP_{dc}/di_d = 0$, the maximum power transfer to the DC would be:

$$\frac{dP_{dc}}{di_d} = \frac{3}{2}E_{\mathrm{m}} - 3Ri_d = 0 \rightarrow i_d = i_{d\_\max} = \frac{E_{\mathrm{m}}}{2\left(R + R_{\mathrm{SDBR}}\right)} \tag{10.21}$$

Substituting Equation (10.21) into (10.20):

$$P_{dc\_\max} = \frac{3E_{\mathrm{m}}^2}{8\left(R + R_{\mathrm{SDBR}}\right)} \tag{10.22}$$

And the voltage source converter operation is possible when

$$P_{dc} \leq P_{dc\_\max} \rightarrow \left(\frac{E_{\mathrm{m}}}{U_{\mathrm{dc}}}\right)^2 - \frac{8\left(R + R_{\mathrm{SDBR}}\right)}{3R_{\mathrm{L}}} \geq 0 \tag{10.23}$$

The grid input maximal power $P_{s\_\max}$ can be obtained by putting Equation (10.22) into Equation (10.8) for Q_s. Thus:

$$P_{s\_\max} = \frac{3E_{\mathrm{m}}^2}{4\left(R + R_{\mathrm{SDBR}}\right)} \tag{10.24}$$

In light of the above analysis, maximum power transfer of the PMSG GSC, during fault would be mitigated, reducing the total current and oscillations by employing the topology of SDBR.

10.2.2   PMSG WIND TURBINE WITH BFCL

The control structures of a BFCL are shown in Figure 10.3, and it is basically made up of two distinctive parts, as earlier discussed in Chapter 9. The BFCL main part is

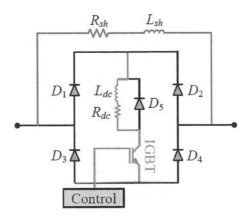

FIGURE 10.3 Control structure of BFCL in PMSG wind turbine.

a typical bridge circuit with four diodes (D_1–D_4), while the shunt path made up of inductor (L_{sh}) and resistor (R_{sh}) in series forms the other part of the BFCL circuit. An IGBT switch is connected in series with an inductor (L_{dc}), and (R_{dc}) act as an intrinsic resistance of (L_{dc}) with very small magnitude that is negligible. In the BFCL, the (L_{dc}) inductor is a DC reactor due to the fact that current flows in one direction only through it during positive and negative half cycles of the alternating current. There is a free-wheeling diode D5 that is connected to the DC reactor to protect the system from inductive kick during transient state [21]. The working principle of the BFCL is such that during normal or steady state, the current flows through the *D1-Ldc-Rdc*-IGBT-*D₄* path for the positive half cycle and through the *D3*-IGBT-*Rdc-Ldc-D₂* path for the negative half cycle. It should be noted that the shunt path of the BFCL has a very high impedance, making the bridge switch carry the line current and some negligible leakage currents [20, 22]. The control strategy of the BFCL used in this chapter is based on the threshold grid voltage, which is the same for the SDBR control strategy for fair comparison. The parameters of the BFCL were already given in Table 10.2.

10.2.3 PMSG WIND TURBINE WITH THE CBFCL

The CBFCL circuit has four diodes with a switching circuitry of a DC reactor (L_D) and (r_D), as shown in Figure 10.4. The shunt path is made up of a capacitor C_{sh} with a series resistor R_{sh}. In addition, there are two fast recovery diodes (D_5 and D_6) in the bridge circuit. The parameters of the CBFCL are shown in Table 10.2 and the switching strategy is the same as those of the SDBR and BFCL, for effective comparative study. For practical realization of the operation of the capacitor in a high voltage, the control input which is the duty cycle of a Pulse Width Modulator (PWM), is a function of V_C, V_S, V_L in the equivalent circuit of Figure 10.5. The generated pulses from the PWM signal generator are used to drive the IGBTs so that the fault current could be suppressed. The mathematical dynamic model of the CBFCL in PMSG wind turbine based on on-state (normal or steady state operation) and off-state (grid fault scenario) is described as follows.

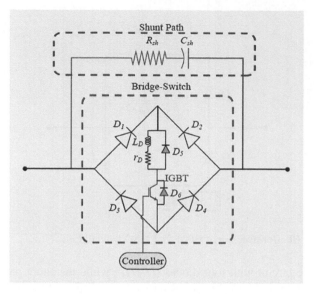

FIGURE 10.4 Control structure of CBFCL in PMSG wind turbine.

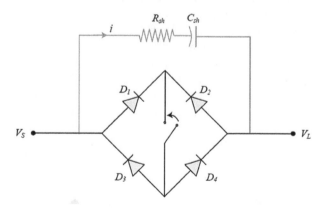

FIGURE 10.5 On-state Equivalent CBFCL in PMSG wind turbine.

The on-state equation is based on the Kirchoff's voltage law being applied to the terminals of the equivalent circuit of the bridge in Figure 10.4, shown in Figure 10.5. Thus:

$$C_{sh} \frac{dV_C}{dt} = i \tag{10.25}$$

$$i = \frac{V_s - V_c - V_L}{R_{sh}} \tag{10.26}$$

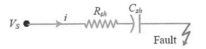

FIGURE 10.6 Off-state Equivalent CBFCL in PMSG wind turbine.

$$C_{sh}\frac{dV_C}{dt} = \frac{V_s - V_c - V_L}{R_{sh}} \tag{10.27}$$

Simplification of Equation (10.27) leads to:

$$\frac{dV_C}{dt} = -\frac{V_c}{R_{sh}C_{sh}} + \frac{V_s}{R_{sh}C_{sh}} - \frac{V_L}{R_{sh}C_{sh}} \tag{10.28}$$

where V_C, C, i, R_{sh}, V_s, V_L, are the capacitor's voltage, capacitance of the capacitor, current in the shunt path, resistance of the shunt path, supply voltage and load voltage, respectively.

The off-state equation is based on Kirchoff's voltage law being applied to the terminals of the equivalent circuit of the bridge in Figure 10.4, shown in Figure 10.6. Thus:

$$C_{sh}\frac{dV_C}{dt} = i \tag{10.29}$$

The current during fault is expressed based on Ohm's law as:

$$i = \frac{V_s - V_c}{R_{sh}} \tag{10.30}$$

$$C_{sh}\frac{dV_C}{dt} = \frac{V_s - V_c}{R_{sh}} \tag{10.31}$$

Simplification of Equation (10.31) leads to:

$$\frac{dV_C}{dt} = -\frac{V_c}{R_{sh}C_{sh}} + \frac{V_s}{R_{sh}C_{sh}} \tag{10.32}$$

10.3 CONTROL STRATEGY OF THE PMSG WIND TURBINE

The control structure of the PMSG wind turbine is shown in Figure 10.7, where the full power converter is used for isolation of the wind generator from the power network, for better protection during grid fault. This is because the grid faults have huge impact on the direct drive wind energy conversion technology. The MSC regulates the active and reactive power of the PMSG by carrying out abc to dq transformation using angle position rotor (θ_r) computed from the rotor speed. The d-axis and q-axis currents (I_{sd}), (I_{sq}) control active power (P_s) and reactive power (Q_s) of the PMSG

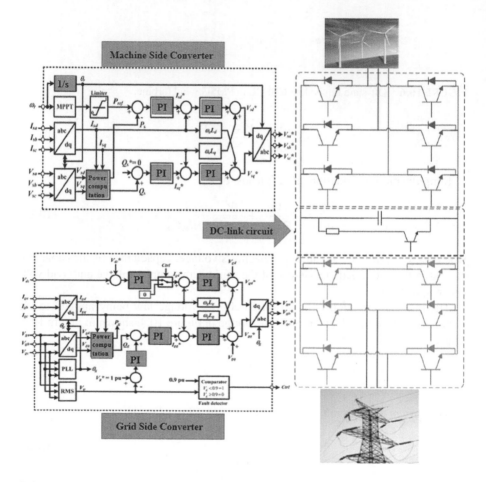

FIGURE 10.7   Control strategy of the PMSG wind turbine.

wind turbine. The reference active power (P_{ref}) is derived from the MPPT of the wind turbine characteristics earlier discussed, while the reference reactive power (Q_s^*) is fixed at 0, for unity power factor. (V_{sa}^*, V_{sb}^*, V_{sc}^*) are generated as reference voltages switching, considering references V_{sd}^* and V_{sq}^* voltages.

The GSC control considers the d-q rotating reference frame and the voltage of the power grid along with the speed of rotation. (I_{ga}, I_{gb}, I_{gc}) three-phase currents and (V_{ga}, V_{gb}, V_{gc}) three-phase voltages are converted to their rotating reference d-q frame. The phase angle (θ_g) on the GSC is obtained from the Phase Locked Loop (PLL) structure. For effective grid voltage transformation, V_{gd} is adjusted to a constant and V_{gq} to zero, in the stationary reference frame and the rotating reference d-q frame. The d-axis current (I_{gd}) and the q-axis current (I_{gq}), regulate the active and reactive power that the PMSG is dissipating to the grid. V_{gd}^* and V_{gq}^* are transformed to (V_{ga}^*, V_{gb}^*, V_{gc}^*) and used for switching purpose. The DC-link voltage (V_{dc}) is usually kept at unity for effective active power transfer. The DC-link determines the d-axis current (I_{gd}^*) reference signal, while the reactive power determines the q-axis current (I_{gq}^*) reference

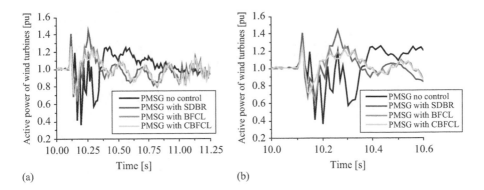

(a) (b)

FIGURE 10.8 (a) Active power of the PMSG wind turbine. (b) Zoom of Figure 10.8(a).

signal. The voltage is proportional to the reactive power, making the terminal wind turbine voltage to be kept at 1.0 pu.

10.4 EVALUATION OF THE SYSTEM PERFORMANCE

The evaluation of the model system of PMSG wind turbine with the FCLS was done using PSCAD/EMTDC [23]. A severe three-phase fault of 100 ms happening at 10.1 s, with the circuit breakers operation sequence opening and reclosing at 10.2 s and 11 s, respectively, was considered in this study. The switching frequency used for the MSC is 1000 Hz, while that of the GSC is 1050 Hz. The solution time step is 10 μS. The evaluation of the system performance was done considering the position of the SDBR, BFCL and CBFCL at the GSC of the PMSG wind turbine [24]. A scenario where no control was implemented in the PMSG wind turbine without considering any of the FCLs was also investigated. The PMSG wind generator was operating at its rated speed during the grid fault. Figures 10.8a–10.12 show the performances of the various variables of the PMSG wind turbine, and the zooms of these figures are shown in Figures 10.8(b)–10.12(b).

Figures 10.8a and 10.9 show the active power and DC-link voltage of the PMSG wind turbine for no FCL control and with SDBR, BFCL and CBFCL control

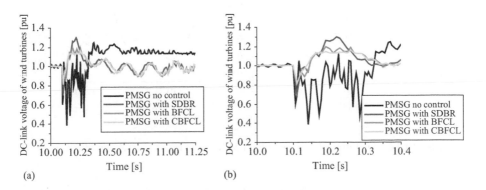

(a) (b)

FIGURE 10.9 (a) DC-link voltage of the PMSG wind turbine. (b) Zoom of Figure (a).

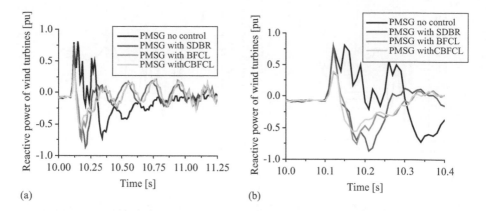

FIGURE 10.10 (a) Reactive power of the PMSG wind turbine. (b) Zoom of Figure (a).

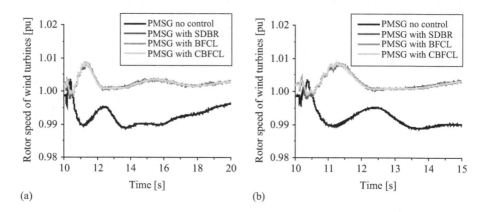

FIGURE 10.11 (a) Rotor speed of the PMSG wind turbine. (b) Zoom of Figure 10.11 (a).

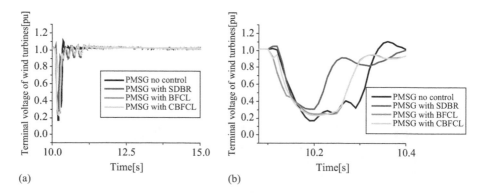

FIGURE 10.12 (a) Terminal voltage of the PMSG wind turbine. (b) Zoom of Figure (a).

strategy. From the responses of these figures, inserting SDBR, BFCL and CBFCL on the GSC of the PMSG wind turbine has major effect on the active power and DC-link voltage during transient state. This is because the PMSG wind turbine is decoupled fully from the power grid using back-to-back power converter. The undershoot, overshoot and settling time of the active power and DC-link voltage are better in Figures 10.8a and 10.9 with the scenarios of the FCLs, compared to when no FCL was employed. Connecting the SDBR on GSC of the PMSG ensures the expected high voltage of the wind generator stator circuitry is divided because of the series connection strategy. However, the performance of the CBFCL is better than those of the SDBR and the BFCL, because of the additional energy buffer from the capacitive circuit of the CBFCL. Figure 10.10 shows that the reactive power was better controlled using the CBFCL than the other FCLs and when no FCL was employed. Due to the capacitive circuit of the CBFCL in the PMSG wind turbine during transient state, the reactive power of the wind turbine would be enhanced.

In Figure 10.11(a), the performance of the rotor speed of the PMSG wind turbine is better with the use of FCLs. The responses of the rotor speed are the same for the FCLs in both steady and transient states. This is because the FCL control technique in the PMSG wind turbine has the ability to improve its mechanical output slightly in steady state and limit its speed during transient state. Therefore, the performance of the rotor speed in Figure 10.11(a) would be with fewer oscillations and faster settling time. Furthermore, because of the ability of the SDBR to boost the reactive power dissipation as shown in Figure 10.10, the terminal voltage of the PMSG wind turbine would be more improved as shown in Figure 10.12. The performance of the SDBR FCL is better than the BFCL and CBFCL for the PMSG wind turbine during transient state. The response of the BFCL and CBFCL are the same for the terminal voltage of the PMSG during transient, though with faster settling time than the SDBR FCL. Table 10.3 shows the numerical index performance of the different FCLs based on the presented simulation results. In general, the use of the FCLs in the PMSG wind turbine would result in no power converter loss in control, with little or no induced overvoltage. The FCLs topology would also reduce high current flow, leading to no dangerous overvoltage and excessive charging current in the power converter's DC-link capacitor. Although the PMSG wind turbines are more expensive than the DFIG and Squirrel Cage Induction Generator (SCIG) wind turbines; however, they have better FRT or LVRT performances. By adding additional FCL protection topologies, to the PMSG, the overall cost would not be marginally high, due to the fact that the FCLs are cheap switching devices. The advancements in power electronic technologies would further drastically reduce the cost of the FCLs embedded in the PMSG wind turbines. As part of future work, the proportional integral controllers for PMSG wind turbines would be replaced by the dragon fly optimization algorithm.

TABLE 10.3
Numerical Index Performance of the Fault Current Limiters

PMSG variables	Metrics of evaluation	No control	SDBR	BFCL	CBFCL
Active power	Overshoot	1.38 pu	1.40 pu	1.20 pu	1.20 pu
	Settling time	1.00 s	0.60 s	0.60 s	0.60 s
	Dip	0.40 pu	0.62 pu	0.62 pu	0.70 pu
DC-link voltage	Overshoot	1.20 pu	1.30 pu	1.20 pu	1.10 pu
	Settling time	1.25 s	0.40 s	0.40 s	0.40 s
	Dip	0.40 pu	0.70 pu	0.80 pu	0.80 s
Reactive power	Overshoot	0.75 pu	0.75 pu	0.75 pu	0.75 pu
	Settling time	1.10 s	0.50 s	0.40 s	0.40 s
	Dip	−0.80 pu	−0.80 pu	−0.75 pu	−0.50 pu
Rotor speed	Overshoot	1.03 pu	1.01 pu	1.01 pu	1.01 pu
	Settling time	15.00 s	2.00 s	2.00 s	2.00 s
	Dip	0.99 pu	1.00 pu	1.00 pu	1.00 pu
Terminal voltage	Overshoot	1.20 pu	1.05 pu	1.00 pu	1.00 pu
	Settling time	0.45 s	0.45 s	0.40 s	0.40 s
	Dip	0.18 pu	0.30 pu	0.20 pu	0.20 pu

10.5   CHAPTER CONCLUSION

The improved performances of the PMSG wind turbine considering FCLs based on SDBR, bridge-type FCL and CBFCL were investigated in this chapter. The same threshold value of the power grid voltage during transient state was used as the switching strategy for the considered FCLs, for effective comparison. The responses of the FCLs were investigated considering a severe three-phase-to-ground fault at the terminals of the PMSG. A scenario where no FCL was employed in the PMSG was also investigated. From the obtained results, when no FCL was implemented, the PMSG wind turbine experienced substantial consequences during fault. The use of the FCLs improved the performance of the PMSG wind turbine. However, CBFCL performance was superior to the SDBR and BFCL under severe fault condition. The CBFCL provides smoother and faster response with better overshoot and fast settling time for most of the PMSG variables than the other FCLs. Though the SDBR FCL performed better in the terminal voltage response of the PMSG wind turbine during fault scenario with regard to faster recovery of the terminal voltage; however, the settling time was lower than those of the BFCL and CBFCL. Therefore, the CBFCL provides a good example of solving and improving the FRT or low voltage ride through capability of PMSG wind farms.

REFERENCES

[1] M. Islam, M. Huda, J. Hasan, M. A. H. Sadi, A. AbuHusscin, T. K. Roy, and M. Mahmud, "Fault ride through capability improvement of DFIG based wind farm using nonlinear controller based bridge-type flux coupling non-superconducting fault current limiter," *Energies*, vol. 13, no. 7, p. 1696, 2020.

[2] M. Firouzi, "Low-voltage ride-through (LVRT) capability enhancement of DFIG-based wind farm by using bridge-type superconducting fault current limiter (BTSFCL)," *Journal of Power Technologies*, vol. 99, no. 4, pp. 245–253, 2020.

[3] M. R. Islam, J. Hasan, M. M. Hasan, M. N. Huda, M. A. H. Sadi, and A. AbuHussein, "Performance improvement of DFIG-based wind farms using NARMA-l2 controlled bridge-type flux coupling non-superconducting fault current limiter," *IET Generation, Transmission & Distribution*, vol. 14, no. 26, pp. 6580–6593, 2021.

[4] J. Hasan, M. R. Islam, M. R. Islam, A. Z. Kouzani, and M. P. Mahmud, "A capacitive bridge-type superconducting fault current limiter to improve the transient performance of DFIG/PV/SG-based hybrid power system," *IEEE Transactions on Applied Superconductivity*, 2021, doi: 10.1109/TASC.2021.3094422.

[5] G. Rashid, and M. H. Ali, "A modified bridge-type fault current limiter for fault ride-through capacity enhancement of fixed speed wind generator," *IEEE Transactions on Energy Conversion*, vol. 29, no. 2, pp. 527–534, 2014.

[6] M. M. Moghimian, M. Radmehr, and M. Firouzi, "Series resonance fault current limiter (SRFCL) with MOV for LVRT enhancement in DFIG-based wind farms," *Electric Power Components and Systems*, vol. 47, no. 19–20, pp. 1814–1825, 2019.

[7] J. Yang, J. E. Fletcher, and J. O'Reilly, "A series-dynamic-resistor-based converter protection scheme for doubly-fed induction generator during various fault conditions," *IEEE Transactions on Energy Conversion*, vol. 25, no. 2, pp. 422–432, 2010.

[8] Z. Din, J. Zhang, Z. Xu, Y. Zhang, and J. Zhao, "Low voltage and high voltage ride-through technologies for doubly fed induction generator system: Comprehensive review and future trends," *IET Renewable Power Generation*, vol. 15, no. 3, pp. 614–630, 2021.

[9] M. Firouzi, and G. Gharehpetian, "Improving fault ride-through capability of fixed-speed wind turbine by using bridge-type fault current limiter," *IEEE Transactions on Energy Conversion*, vol. 28, no. 2, pp. 361–369, 2013.

[10] S. B. Naderi, M. Jafari, and M. T. Hagh, "Parallel-resonance-type fault current limiter," *IEEE Transactions on Industrial Electronics*, vol. 60, no. 7, pp. 2538–2546, 2012.

[11] M. Firouzi, and G. B. Gharehpetian, "LVRT performance enhancement of DFIG-based wind farm by capacitive bridge-type fault current limiter," *IEEE Transactions on Sustainable Energy*, vol. 9, no. 3, pp. 1118–1125, 2017.

[12] M. A. H. Sadi, A. AbuHussein, and M. A. Shoeb, Transient performance improvement of power systems using fuzzy logic controlled capacitive-bridge type fault current limiter, *IEEE Transactions on Power Systems*, vol 36, no 1, pp. 323–335, 2020.

[13] A. Padmaja, A. Shanmukh, S. S. Mendu, R. Devarapalli, J. Serrano Gonz´alez, and F. P. Garc´ıa M´arquez, "Design of capacitive bridge fault current limiter for low-voltage ride through capacity enrichment of doubly fed induction generator-based wind farm," *Sustainability*, vol. 13, no. 12, p. 6656, 2021.

[14] K. E. Okedu, and H. Barghash, "Enhancing the transient state performance of permanent magnet synchronous generator based variable speed wind turbines using power converters excitation parameters," *Frontiers in Energy Research-Smart Grids*, vol. 9, pp. 109–120, Article 655051, 2021, doi: 10.3389/fenrg.2021.655051.

[15] K. E. Okedu, and S.M. Muyeen, "Enhanced performance of PMSG wind turbines during grid disturbance at different network strengths considering fault current limiter," *International Transactions on Electrical Energy Systems, Wiley*, vol. e12985, no. 6, pp. 1–21, 2021, doi: 10.1002/2050-7038.12985.

[16] K. E. Okedu, S. M. Muyeen, R. Takahashi, and J. Tamura, "Wind farms fault ride through using DFIG with new protection scheme," *IEEE Transactions on Sustainable Energy*, vol. 3, no. 2, pp. 242–254, 2012.

[17] K. E. Okedu, "Effect of ECS low pass filter timing on grid frequency dynamics of a power network considering wind energy penetration," *IET Renewable Power Generation*, vol. 11, no. 9, pp. 1194–1199, 2017.

[18] K. E. Okedu, "Determination of the most effective switching signal and position of braking resistor in DFIG wind turbine under transient conditions," *Electrical Engineering*, vol. 102, no. 11, pp. 471–480, 2020.

[19] K. E. Okedu, S. M. Muyeen, R. Takahashi, and J. Tamura, "Wind farm stabilization by using DFIG with current controlled voltage source converters taking grid codes into consideration," *IEEJ Transactions on Power and Energy*, vol. 132, no. 3. pp. 251–259, 2012.

[20] G. Rashid, and M. H. Ali, "Fault ride through capability improvement of DFIG based wind farm by fuzzy logic controlled parallel resonance fault current limiter," *Electric Power Systems Research*, vol. 146, pp. 1–8, 2017.

[21] M. R. Islam, M. N. Huda, J. Hasan, M. A. H. Sadi, A. Abuhussein, T. K. Roy, and M. P. Mahmud, "Fault ride through capability improvement of DFIG based wind farm using nonlinear controller based bridge-type flux coupling non-superconducting fault current limiter," *Energies*, vol. 13, p. 1696, 2020, doi: 10.3390/en13071696

[22] G. Rashid, and M. H. Ali, "Nonlinear control-based modified BFCL for LVRT capacity enhancement of DFIG-based wind farm," *IEEE Transactions on Energy Conversion*, vol. 32, pp. 284–295, 2016.

[23] PSCAD/EMTDC Manual; Version 4.6.0; Manitoba HVDC Lab.: Winnipeg, MB, Canada, 2016.

[24] K. E. Okedu, "Improving the performance of PMSG wind turbines during grid fault considering different strategies of fault current limiters," *Frontiers in Energy Research-Smart Grids*, vol. 10, no. 909044, pp. 1–12, 2022, doi: 10.3389/fenrg.2022.909044.

11 Comparative Study of DFIG and PMSG with Different Fault Current Limiters

LIST OF SYMBOLS

X	reactance, Ω
R	resistance, Ω
Z	impedance, Ω
T_m	torque, N
ρ	air density, kg/m^3
R	radius, m
V_w	wind speed, m/s^2
$C_p(\lambda, \beta)$	power coefficient
λ	is the ratio of the tip speed
C_t	is the turbine coefficient
P_s	is the stator power, Watts
P_r	is the rotor power, Watts
i_{qr}	quadrature axis rotor current, A
i_{dr}	direct axis rotor current, A
Q_s	stator reactive power, VA
L_s	stator inductance, H
L_m	magnetizing inductance, H
L_r	rotor inductance, H
φ_s	stator flux, T
ω_s	stator angular frequency, Hz
ω_r	rotor angular frequency, Hz
σ	rotor leakage factor
α, β	stationary frames
r, s	DFIG rotor and stator quantities
g	DFIG grid-side converter circuit quantity
L	inductance, H
R	resistance, Ω
V_{dc}	dc-link voltage, V
P_{ref}	reference power of turbine, W
θ_r	rotor angle position

DOI: 10.1201/9781003350910-11

I_{sd}, I_{sq}	direct and quadrature stator current, A
$V_{sa}^*, V_{sb}^*, V_{sc}^*$	reference abc stator voltages, V
V_{sd}^* and V_{sq}^*	reference dq stator voltages, V
I_{ga}, I_{gb}, I_{gc}	abc grid currents, A
V_{ga}, V_{gb}, V_{GC}	abc grid voltages, V
V_{s+}, V_{s-}	components of the stator's voltage positive and negative sequences, V
τ_s	time constant of the stator flux, S
ψ_{sn}	natural flux, T
$\bar{v}_{ro}$	rotor-induced voltage
i_d, i_q	dq current input of the rectifier's axes, A
e_d, e_q	dq voltage of the grid voltage axes components, V
ω	angular frequency voltage, Hz
β_d, β_q	modulating signal of the rectifier's d and q axes components
r	modulation signal vector norm
E_m	phase grid voltage amplitude, V
R_{SDBR}	resistance of the series dynamic braking resistor, Ω
R_L	resistance of the load, Ω
P_{dc}	available power at the DC, W
P_{dc_max}	maximal available power at the DC, W

11.1   CHAPTER INTRODUCTION

As the amount of wind energy penetration into existing power grids is increasing by the day, with an average forecast of 75 GW per year over the 2021–2026 period [1], it is very vital to learn new methods of stabilizing the power grids, for smooth operation [1, 2]. The grid requirements that are recently emerging as a guild to operate wind farms require robust voltage and frequency controls. The technology of variable speed wind turbines is mostly employed because of the extensive range of wind speed operation [3]. For wind energy conversion, the Doubly Fed Induction Generator (DFIG) and the Permanent Magnet Synchronous Generator (PMSG) are the two basic variable speed wind turbines usually used in modern wind farms. The revolution in power electronics and drives in control mechanisms has contributed greatly to the advancements of wind turbines from fixed speed to variable speed technology [4, 5]. Simplicity of operation, rugged construction, low cost and little maintenance are some of the merits of the fixed speed wind turbines. However, the huge requirement of large reactive during grid disturbance, in order to survive recovery of air gap flux and no control for voltage and frequency are some of the major concerns of this class of wind turbines that make them limited in wind energy applications. Consequently, modern wind farms are built using variable speed wind turbines because of their high energy capture efficiency, good voltage control and reduced mechanical drive train stresses [6]. The technologies of the DFIG and PMSG wind turbines have power converters that are connected back to back. While the DFIG has a gearbox system and a power converter rating of 20–30%, the PMSG has high initial cost because of its fully rated power converters.

The technology of the DFIG is such a way that Rotor Side Converter (RSC), otherwise known as the Machine Side Converter (MSC), and the Stator Side

Converter (SSC), otherwise known as the Grid Side Converter (GSC), is connected between the DC-link voltage, for easy regulation of active and reactive power. For effective energy capture, this class of wind turbine operates in a wide range [7, 8] and has a good pitch control mechanism that helps rebuild its voltage after grid disturbance [9, 10].

The technology of the PMSG wind turbine is such a way that its back-to-back power converter is fully rated, unlike the DFIG wind turbine technology that is partially rated. Consequently, the maximum flexibility and better control of real and reactive power are more likely in this class of wind turbine. Though, the high initial cost of the PMSG is quite discouraging

The separate control strategies of DFIG and PMSG wind turbines using various schemes already exist in the literature, ranging from Fault Current Limiters (FCLs) in the DFIG wind turbines [11–14], reactive power compensation, crowbar and DC chopper [15, 16] and sliding mode control for Maximum Power Point Tracking (MPPT) [17, 18]. The Fault Ride Through (FRT) assessment of a DFIG wind turbine was carried out in [19–21], with the help of different control topologies, and algorithms for wind energy conversion were reported in [22]. On the other hand, for the PMSG wind turbine, the limitation of the maximum current, and MPPT power converters were studied in [23, 24], where the DC-link voltage was kept constant near its limit during grid fault. The enhancement of PMSG wind turbine was also done in [25], using the Superconducting Fault Current Limiter (SFCL) control strategy.

The hardware-based solutions of FCLs in DFIG and PMSG wind turbines in wind farms have shown promising results in augmenting the performance of these wind turbines during grid faults [26]. There are two types of FCLs: SFCL having no power loss, with good speed control, though complex circuitries [27–29] and Non-superconducting Fault Current Limiter (NSFCL) that can resolve the issues of the SFCLs, in addition to providing FRT capabilities [30, 31]. The Insulated Gate Bipolar Transistor (IGBT) technology led to the wider use of NSFCLs than the others. The Series Dynamic Braking Resistors (SDBRs) [32, 33] and Bridge Fault Current Limiters (BFCLs) having resistive, inductive and capacitive elements [34] and Parallel Resonance FCL (PRFCL) [35] have been employed in variable speed wind turbines FRT.

In this chapter, the augmentation analyses of the DFIG and PMSG wind turbines regarding stability issues experienced by DFIG and PMSG wind turbines were investigated considering various existing FCLs. The details of the wind turbine modeling characteristics of both wind generators were presented along with their control topologies. Both wind turbines were subjected to a severe bolted three-line-to-ground fault, without any protection or enhancement scheme of the FCL, to test the robustness of the controllers. The mathematical dynamics of inserting SDBR, BFCL and CBFCL at the stator of both wind turbines considering the same condition of operation, for fair comparison were also presented. The same effective size of the SDBR and the same parameters of the BFCL and CBFCL were used for both wind turbine technologies. The grid voltage was used as the switching signal, during grid fault. There are a limited number of reports in the literature that considered the scenarios of these FCLs in both wind turbines. Most reports in the literature considered these scenarios on a separate basis FRT enhancement of both wind generators. The study was carried out using Power System Computer Aided Design and Electromagnetic Transient Including DC (PSCAD/EMTDC) environment [36].

11.2   MODELING AND CONTROL

The details of the DFIG and PMSG wind turbine modeling and characteristics have already been presented in earlier chapters of this book (Chapters 1–10). Readers should refer to these chapters for more details.

11.3   DFIG AND PMSG MODEL SYSTEMS WITH THE FAULT CURRENT LIMITERS

The model systems of the DFIG and PMSG wind turbines with the three FCLs are shown in Figures 11.1 and 11.2. The parameters of the wind turbines and those of the FCLs are given in Tables 11.1 and 11.2. The mathematical dynamics of the three FCLs are presented in the subsequent section of this chapter.

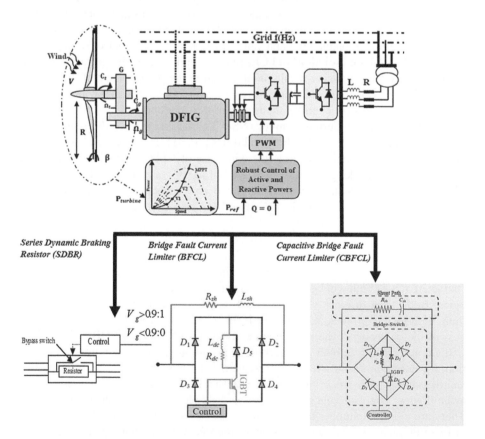

FIGURE 11.1   DFIG wind turbine model with the fault current limiters.

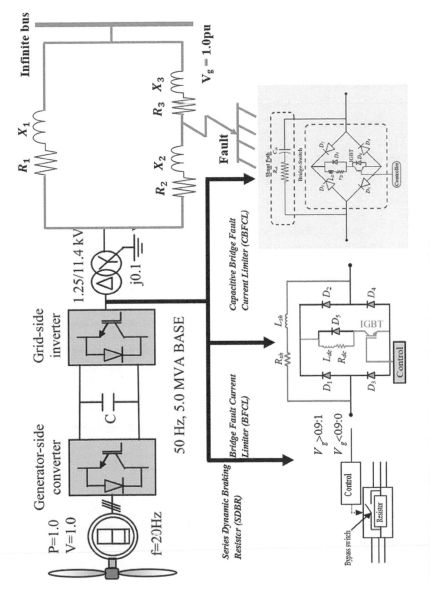

FIGURE 11.2   PMSG wind turbine model with the fault current limiters.

TABLE 11.1

Rating of Parameters of the Wind Turbines

DFIG wind turbine		PMSG wind turbine	
Rated power	5.0 MW	Rated power	5.0 MW
Stator resistance	0.01 pu	Stator resistance	0.01 pu
d-axis reactance	1.0 pu	d-axis reactance	1.0 pu
q-axis reactance	0.7 pu	q-axis reactance	0.7 pu
Machine inertia (H)	3.0	Machine inertia (H)	3.0
Effective DC-link protection	0.2 Ω	Effective DC-link protection	0.2 Ω

TABLE 11.2

Parameters of the Fault Current Limiters

SDBR		BFCL				CBFCL			
		R_{sh}	L_{sh}	R_{dc}	L_{dc}	R_{sh}	L_{sh}	R_{dc}	C_{sh}
Series resistance (R_s) 0.1 pu		20 Ω	250 mH	0.003 Ω	1 mH	20 Ω	250 mH	0.003 Ω	69 μF

11.4 MATHEMATICAL DYNAMICS OF SDBR IN DFIG AND PMSG WIND TURBINES

11.4.1 SDBR IN DFIG WIND TURBINES

The dynamics of the SDBR in DFIG wind turbines are represented in Figure 11.3. The stator voltage of the DFIG wind turbine during transient conditions [37] is.

$$v_s^s = V_{s+}e^{j\omega_s t} + V_{s-}e^{-j\omega_s t} \tag{11.1}$$

V_{s+}, V_{s-} are components of the stator's voltage positive and negative sequences. At normal state, the stator flux is:

$$\psi_{ss}^s = \frac{v_{s+}e^{j\omega_s t}}{j\omega_s} + \frac{v_{s-}e^{-j\omega_s t}}{-j\omega_s} \tag{11.2}$$

With a sudden drop in grid voltage, there will be transient components in the stator flux to counteract transition in the state variables [38]. Therefore, the stator flux would have the natural flux (ψ_{sn}) expressed as:

$$\psi_s^s = \psi_{ss}^s \psi_{sn}^s = \frac{v_{s+}e^{j\omega_s t}}{j\omega_s} + \frac{v_{s-}e^{-j\omega_s t}}{-j\omega_s} + \psi_{sn}^s e^{\frac{-t}{\tau_s}} \tag{11.3}$$

$\tau_s = \dfrac{L_s}{R_s}$ is the time constant of the stator flux.

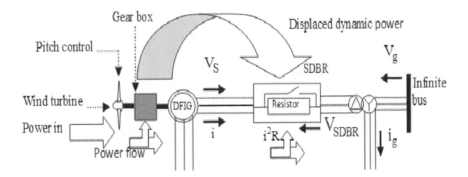

FIGURE 11.3 Dynamics of SDBR in DFIG wind turbines.

During transient state, the forced flux and the natural flux (ψ_{sn}) in Equation (11.3) occur, which are the first and second terms. The rotor reference frame is related to the stator flux by

$$\psi_s^r = \psi_s^s e^{-j\omega_r t} \tag{11.4}$$

While the rotor-induced voltage is

$$\bar{v}_{ro} = \frac{L_m}{L_s} \frac{d\bar{\psi}_s}{dt} \tag{11.5}$$

The open circuit rotor voltage of the DFIG wind turbine is obtained from Equations (11.3) to (11.5) as:

$$\bar{v}_{ro} = \frac{L_m}{L_s} s V_+ e^{js\omega_s t} + \frac{L_m}{L_s}(s-2)V_- e^{j(2-s)\omega_s t} + \frac{L_m}{L_s}\left(j\omega_r + \frac{1}{\tau_s}\right)\psi_{sn} e^{-\frac{t}{\tau_s}}e^{j\omega_r t} \tag{11.6}$$

Neglecting the $\dfrac{1}{\tau_s}$ term leads to in Equation (11.6) gives

$$\bar{v}_{ro} = \frac{L_m}{L_s} s V_+ e^{js\omega_s t} + \frac{L_m}{L_s}(s-2)V_- e^{j(2-s)\omega_s t} + \frac{L_m}{L_s} A(1-s)e^{-\frac{t}{\tau_s}}e^{j\omega_r t} \tag{11.7}$$

During grid fault, the DFIG stator flux is comprised of forced component and flux natural, with high rotor voltage transient and natural flux component. However, there is decay of the natural flux with time constant $\tau_s = \dfrac{R_s}{L_s}$ of the stator circuit. Consequently, the stator resistance would increase due to the SDBR as below:

$$R_{seffective} = R_s + R_{sdbr} \tag{11.8}$$

The new time constant would now be $\tau_{\text{seffective}} = \dfrac{L_s}{R_{\text{seffective}}}$, making the total current lower, with reduced oscillations during transient state.

11.4.2 SDBR in PMSG Wind Turbines

The dynamics of the SDBR in DFIG wind turbines are represented in Figure 11.4. Based on Park's transformation, the PMSG grid-connected voltage source converter could be modeled in rotating frame. The mathematical model for the balanced three-phase voltage source converter in Figure 11.2 is given as [39]:

$$e_d = -\omega L i_q + L\frac{di_d}{dt} + \left(R + R_{\text{SDBR}}\right)i_d + 0.5U_{dc}\beta_d \tag{11.9}$$

$$e_q = -\omega L i_d + L\frac{di_q}{dt} + \left(R + R_{\text{SDBR}}\right)i_q + 0.5U_{dc}\beta_q \tag{11.10}$$

$$C\frac{dU_{dc}}{dt} = 0.75\left(i_d\beta_d + i_q\beta_q\right) - \frac{U_{dc}}{R_L} \tag{11.11}$$

$$r = \sqrt{\beta_d^2 + \beta_q^2} \tag{11.12}$$

From Equations (11.9) to (11.12), i_d, i_q represent the dq current input of the rectifier's axes, e_d, e_q are known as the dq voltage of the grid voltage axes components, ω is the angular frequency voltage, β_d, β_q represent the modulating signal of the rectifier's d and q axes components, while r is the modulation signal vector norm. Considering three-phase transformation based on Park's principle, and that the phase-A grid voltage is in alignment with the dq reference synchronous frame, the source voltage dq components are given as:

$$e_d = E_m \tag{11.13}$$

$$e_q = 0 \tag{11.14}$$

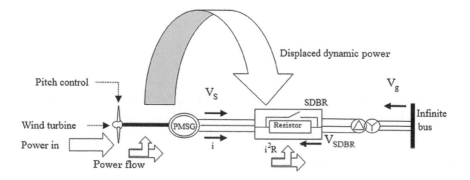

FIGURE 11.4 Dynamics of SDBR in PMSG wind turbines.

From Equation (11.13), E_m is the phase grid voltage amplitude. Consequently, the fed active and reactive rectifier's powers are computed by

$$P_s = \frac{3}{2} E_m i_d \tag{11.15}$$

$$Q_s = -\frac{3}{2} E_m i_q \tag{11.16}$$

To achieve power factor in unity mode of operation, i_{q_ref} can be set to 0. Therefore, for the current regulation to be ideal, $i_q = i_{qref} = 0$. With $i_q = 0$ and R_L, the mathematical model of the voltage source converter under unity power factor can be expressed with the following set of equations:

$$E_m = L \frac{di_d}{dt} + (R + R_{SDBR}) i_d + 0.5 U_{dc} \beta_d \tag{11.17}$$

$$\beta_q = -\frac{2\omega L}{U_{dc}} i_q \tag{11.18}$$

$$C \frac{dU_{dc}}{dt} = \frac{3}{4} i_d \beta_d - \frac{U_{dc}}{R_L} \tag{11.19}$$

Equation (11.18) implies in order to ensure operation of unity power factor of the voltage source converter, the component of U_{dc}, should proportionally vary with β_d current. From Equations (11.17) and (11.19), the capacity charge is manipulated by β_d, via the i_d current of the input.

The insertion of the SDBR resistance during transient state in the PMSG converter would affect the maximal power flow and the DC output voltage. During normal condition, the derivative operator relating all terms in Equations (11.17)–(11.19), would be zero. Thus, the new set of equations would be:

$$E_m = (R + R_{SDBR}) i_d + 0.5 U_{dc} \beta_d \tag{11.20}$$

$$\beta_q = -\frac{2\omega L}{U_{dc}} i_q \tag{11.21}$$

$$i_d = \frac{4 U_{dc}}{3 \beta_d R_L} \tag{11.22}$$

Putting Equation (11.21) into (11.22), for a given load of R_L, and voltage U_{dc}, the expression of the signal command β_d is

$$6 E_m R_L \beta_d - 8 (R + R_{SDBR}) U_{dc} - 3 R_L \beta_d^2 U_{dc} = 0, \text{ for } \beta_d \neq 0 \tag{11.23}$$

There are two derived solutions from Equation (11.23):

$$\beta_{d1} = \frac{E_m}{U_{dc}} - \sqrt{\left(\frac{E_m}{U_{dc}}\right)^2 - \frac{8(R + R_{SDBR})}{3R_L}} \tag{11.24}$$

$$\beta_{d2} = \frac{E_m}{U_{dc}} + \sqrt{\left(\frac{E_m}{U_{dc}}\right)^2 - \frac{8(R + R_{SDBR})}{3R_L}} \tag{11.25}$$

Since β_d component in Equation (11.24) has very low values, this solution is not admissible. Therefore, the solution of Equation (11.25) is the acceptable solution, making $\beta_d = \beta_{d2}$. And β_d would exist based on the following condition being satisfied:

$$\left(\frac{E_m}{U_{dc}}\right)^2 - \frac{8(R + R_{SDBR})}{3R_L} \geq 0 \tag{11.26}$$

The above condition is with respect to power consumption by the load. The operation of the converter is possible if

$$P_{dc} \leq P_{dc\_max} \tag{11.27}$$

where P_{dc_max} is the maximal available power at the DC-connected side power of the PMSG converter. Considering the power conservation principle, P_{s_max} can be expressed as

$$P_{dc} = \frac{3}{2} E_m i_d - \frac{3}{2}(R + R_{SDBR}) i_d^2 \tag{11.28}$$

By solving Q_s, the maximum power transfer to the DC side is

$$\frac{dP_{dc}}{di_d} = \frac{3}{2} E_m - 3(R + R_{SDBR}) i_d = 0 \rightarrow i_d = i_{d\_max} = \frac{E_m}{2(R + R_{SDBR})} \tag{11.29}$$

Putting Equation (11.29) into (11.28) yields the maximum DC side power as

$$P_{dc\_max} = \frac{3E_m^2}{8(R + R_{SDBR})} \tag{11.30}$$

The PMSG voltage source converter operation is possible when

$$P_{dc} \leq P_{dc\_max} \rightarrow \left(\frac{E_m}{U_{dc}}\right)^2 - \frac{8(R + R_{SDBR})}{3R_L} \geq 0 \tag{11.31}$$

Equation (11.31) reflects the condition of Equation (11.27). Finally, the grid input maximal power P_{s_max} can be obtained by putting Equation (11.30) into Equation (11.16) for Q_s. Thus:

$$P_{s\_max} = \frac{3E_m^2}{4(R + R_{SDBR})} \tag{11.32}$$

Based on Equation (11.32), the maximum power transfer of the GSC of the PMSG during transient state would be reduced by the insertion of SDBR in the GSC. Therefore, the total current reduces its value, thus mitigating the oscillations that normally occur during transient conditions. The SDBR control for both wind turbines is shown in Figure 11.1 and 11.2, considering the threshold grid voltage during grid fault.

11.5 DYNAMICS OF BFCL AND CBFL ON DFIG AND PMSG WIND TURBINES

11.5.1 DFIG AND PMSG WIND TURBINES WITH BFCL

The BFCL control is the same as the SDBR control shown in Figure 11.1 and 11.2, considering the threshold grid voltage during grid fault. Figure 11.5 shows the BFCL architecture, with the main bridge circuit having four diodes (D_1-D_4) and a shunt path having inductor (L_{sh}) and resistor (R_{sh}) in series. The inductor (L_{dc}) is connected to an IGBT switch. (L_{dc}) inductor DC reactor having current flow in only one direction for positive and negative half cycles of AC. D5 is a free-wheeling diode used for protection of inductive kick when there is grid disturbance [40]. In a steady state, for positive half cycle, the current flows through the $D1$-L_{dc}-R_{dc}-IGBT-D_4 path and for negative half cycle, $D3$-IGBT-R_{dc}-L_{dc}-D_2. The impedance of the shunt path is very high, so the line current flows through the bridge switch with some negligible leakage currents [41, 42].

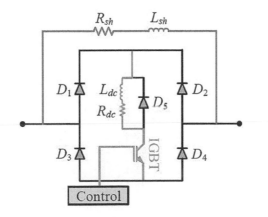

FIGURE 11.5 Control structure of BFCL in PMSG wind turbine.

11.5.2   DFIG AND PMSG WIND TURBINES WITH CBFCL

The switching strategy of the CBFCL is the same as the SDBR and BFCL. Figure 11.6 shows the CBFCL architecture with four diodes, DC reactor (L_D) and (r_D) switching circuitries and shunt path having C_{sh} and series resistor R_{sh}. (D_5 and D_6) act as fast recovery diodes. The following mathematical dynamics would help understand the behavior of the CBFCL in DFIG and PMSG wind turbines during normal and transient states.

From Kirchoff's voltage law, the on-state equations could be derived by applying this law to the equivalent circuit of the bridge in Figure 11.6, shown in Figure 11.7. Thus:

$$C_{sh}\frac{dV_C}{dt} = i \tag{11.33}$$

$$i = \frac{V_s - V_c - V_L}{R_{sh}} \tag{11.34}$$

$$C_{sh}\frac{dV_C}{dt} = \frac{V_s - V_c - V_L}{R_{sh}} \tag{11.35}$$

Equation (11.35) leads to:

$$\frac{dV_C}{dt} = -\frac{V_c}{R_{sh}C_{sh}} + \frac{V_s}{R_{sh}C_{sh}} - \frac{V_L}{R_{sh}C_{sh}} \tag{11.36}$$

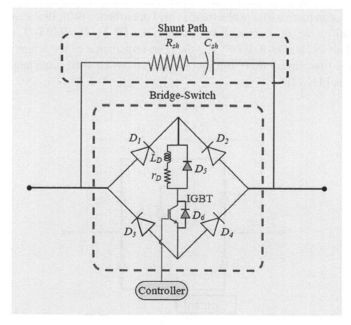

FIGURE 11.6   Control structure of CBFCL in PMSG.

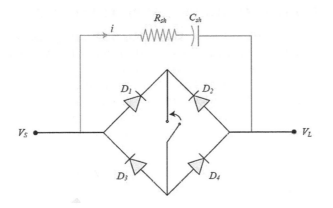

FIGURE 11.7 On-state equivalent CBFCL in PMSG.

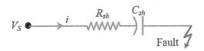

FIGURE 11.8 Off-state equivalent CBFCL in PMSG.

From Equation (11.33) to (11.36), V_C, C, i, R_{sh}, V_s, V_L are the capacitor's voltage, capacitance of the capacitor, current in the shunt path, resistance of the shunt path, supply voltage and load voltage, respectively. On the other hand, the equations for the off-state are based on applying Kirchoff's voltage law to the equivalent circuit of the bridge in Figure 11.6, shown in Figure 11.8. Thus:

$$C_{sh} \frac{dV_C}{dt} = i \tag{11.37}$$

The current during fault is expressed based on Ohm's law as:

$$i = \frac{V_s - V_c}{R_{sh}} \tag{11.38}$$

$$C_{sh} \frac{dV_C}{dt} = \frac{V_s - V_c}{R_{sh}} \tag{11.39}$$

Simplifying of Equation (11.39) leads to:

$$\frac{dV_C}{dt} = -\frac{V_c}{R_{sh}C_{sh}} + \frac{V_s}{R_{sh}C_{sh}} \tag{11.40}$$

11.6   EVALUATION OF THE SYSTEM PERFORMANCE

Efforts were made to run simulation studies for both wind generators in order to compare their FRT characteristics using the different FCLs. A severe three-phase to-ground fault of 100 ms happening at 10.1 s, with the circuit breakers operation sequence opening and reclosing at 10.2 s and 11 s, respectively, on the faulted line close to the terminals of the DFIG and PMSG wind turbines, was investigated. The fault performance with and without considering stability augmentation tools of the SDBR, BFCL and CBFCL are presented below in detail [43].

11.6.1   PERFORMANCE OF THE DFIG AND PMSG WIND TURBINES CONSIDERING SDBR AND BFCL

The DFIG and PMSG wind turbines were subjected to a severe grid fault in the model system considering no insertion of the SDBR and BFCL, and with the insertion of the SDBR and BFCL. Some of the simulation results of the variables of the wind turbines are presented in Figures 11.9–11.13.

From Figure 11.9, the overshoot experienced by DFIG and PMSG DC-link voltage during grid fault could lead to the damage of fragile and vulnerable power

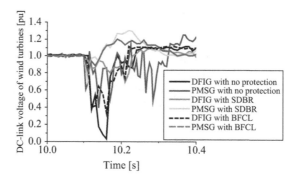

FIGURE 11.9   DC-link voltage of the wind turbines.

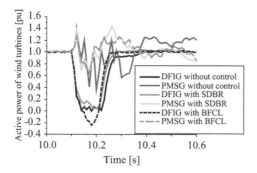

FIGURE 11.10   Active power of the wind turbines.

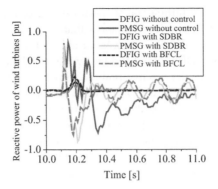

FIGURE 11.11   Reactive power of the wind turbines.

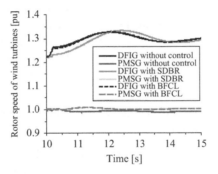

FIGURE 11.12   Rotor speed of the wind turbines.

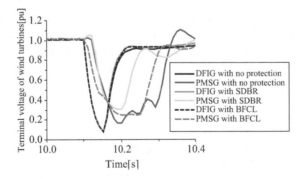

FIGURE 11.13   Terminal voltage of the wind turbines.

converters of the wind turbines, when no protection control strategy is implemented. While the PMSG wind turbine has more overshoot and slower settling time, the DFIG wind turbine has more DC-link voltage dip, with faster settling time. Less voltage dip is achieved with the use of the SDBR in the DFIG than when BFCL was used, while the worst-case scenario was when no control was employed. On the other hand, for the PMSG wind turbine, better performance of the DC-link voltage was achieved using the BFCL with less overshoot than when the SDBR was employed. The same analogy goes for the active power variable in Figure 11.10, with the SDBR performing better in the DFIG wind turbine, while the BFCL in the PMSG wind turbine. The responses for the reactive power and rotor speed are shown in Figures 11.11 and 11.12. The reactive power of the wind turbines was effectively controlled using the SDBR and BFCL, while there was minimal influence of the FCLs on the rotor speed of the wind turbines. The influence of the FCLs is more obvious in the terminal voltage response of the PMSG shown in Figure 11.13, during the grid fault than the DFIG. When no control was implemented in both wind turbines, more over-shoot and less settling time were observed in the variable of the terminal voltage. However, the SDBR gave a quicker recovery of the terminal voltage variable than the BFCL. Based on the above analysis, the use of the FCLs is more obvious for the PMSG than the DFIG because the PMSG-based wind turbine is decoupled fully from the power grid due to its back-to-back power converter. Consequently, the operation of the wind turbines is a critical situation during faulty condition, when no FCL is implemented. In light of this, the subsequent sections of this chapter would consider the investigation of the CBFCL on the two wind turbines separately, in order to know its impact, compared to the SDBR and the BFCL topologies.

11.6.2 Performance of the DFIG Wind Turbine Considering SDBR, BFCL and CBFCL

The performance of the DFIG was further investigated, considering the CBFCL in addition to those obtained using the SDBR and BFCL. Some of the simulation results are presented in Figures 11.14–11.17. In Figure 11.14, the SDBR outperforms the BFCL and the CBFCL, with improved undershoot, overshoot and faster settling time. However, more reactive power was dissipated by the use of the BFCL and the CBFCL in the GSC of the DFIG wind turbine as shown in Figure 11.15. The effect of the FCLs on the terminal voltage is shown in Figures 11.16 and 11.17. In Figures 11.16 and 11.17, when no FCL scheme was considered, the terminal voltage recov-ered slowly, however, with the use of the SDBR, the voltage recovery was faster

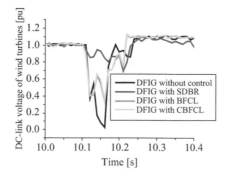

FIGURE 11.14   DC-link voltage of DFIG wind turbine.

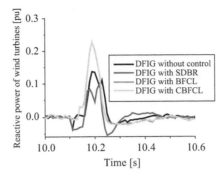

FIGURE 11.15   Reactive power of DFIG wind turbine.

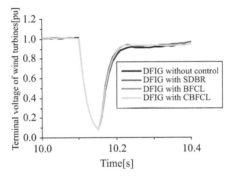

FIGURE 11.16   Terminal voltage of DFIG wind turbine.

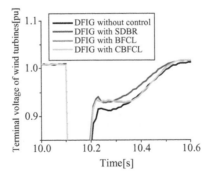

FIGURE 11.17   Terminal voltage of DFIG wind turbine (zoomed).

than the use of the BFCL and CBFCL control strategies. Some of the reasons for the improved performance of the SDBR over the BFCL and the CBFCL are the SDBR has more capability in responding faster during the grid fault by reducing the high current flow, at the same time, ensuring no overvoltage is induced, thus leading to no power converter control loss.

11.6.3 PERFORMANCE OF THE PMSG WIND TURBINE CONSIDERING SDBR, BFCL AND CBFCL

The performance of the PMSG wind turbine was also investigated in this section considering the CBFCL, in addition to those obtained using the SDBR and BFCL. Some of the simulation results are presented in Figures 11.18–11.20 during the grid fault. In Figures 11.18 and 11.19, the CBFCL performed better than the SDBR and BFCL schemes, with improved undershoot, overshoot and faster settling time. However, the SDBR scheme performed better than the BFCL and the CBFCL schemes in the recovery of the terminal voltage, with faster settling time and voltage dip. Some of the reasons for the improved performance of the SDBR scheme in the voltage variable of the PMSG were the same as given for the DFIG wind turbine. Apart from the voltage variable of the PMSG wind turbine, the performance of the CBFCL was better in other variables of the PMSG wind turbine during grid fault.

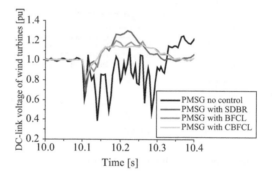

FIGURE 11.18 DC-link voltage of PMSG wind turbine.

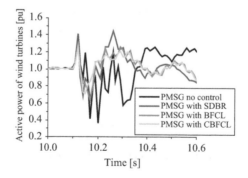

FIGURE 11.19 Active power of PMSG wind turbine.

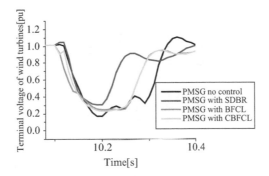

FIGURE 11.20   Terminal voltage of PMSG wind turbine.

Tables 11.3 and 11.4 show the numerical index performance of both wind turbines and the various FCLs employed in this study, where the effect of the FCL was more on the PMSG than the DFIG.

11.7   CHAPTER CONCLUSION

This chapter presented a comparative analysis of the augmentation performance of DFIG and PMSG wind turbines with the same machine ratings, during grid fault considering different FCLs. The considered FCLs were the SDBR, BFCL and the Capacitive Bridge Fault Current Limiter (CBFCL). The wind turbines were operated under the same conditions. Both wind turbines were affected by the connection of the FCLs in their stators during severe grid disturbance. However, the influence of the FCLs was more observed in the PMSG wind turbine compared to the DFIG wind turbine. This is due to the fact that the PMSG wind turbine technology enables the decoupling of its full back-to-back power converters at the power grid. When no control strategy of FCLs was implemented in the wind turbines, critical situation of operation during faulty conditions would result, making the DC-link voltage dip to almost 0.0 pu, with overshoot of 110% and high settling time of over 0.4 s for the DFIG, while for the PMSG an overshoot of 130%, DC-link voltage of 0.4 pu and over 0.5 s of settling time was observed. The performance of the DFIG wind turbine was better using the SDBR, than the use of BFCL and CBFCL control schemes, with improved numerical index performance of 0.7 pu, 0.3 pu and 0.3 pu, for the DFIG and 0.7 pu, 0.8 pu and 0.8 pu, for the PMSG, respectively. Also, faster settling time was also recovered using the FCLs for both wind turbines. On the other hand, the performances of the PMSG variables were better using the CBFCL, than the SDBR and the BFCL control strategies, except for the terminal grid voltage variable. Therefore, it is recommended to use either of the FCLs with DFIG or PMSG-based variable speed wind turbine to get superior FRT performance, based on the intended purpose.

TABLE 11.3

Numerical Index Performance of the Wind Turbines

		DFIG Wind Turbine					PMSG Wind Turbine		
Variable	Metric	No control	SDBR	BFCL	Variable	Metric	No control	SDBR	BFCL
DC-link voltage	Overshoot	1.10 pu	1.08 pu	1.10 pu	DC-link voltage	Overshoot	1.30 pu	1.30 pu	1.20 pu
	Dip	0.01 pu	0.70 pu	0.30 pu		Dip	0.40 pu	0.70 pu	0.80 pu
	Settling time	0.40 s	0.30 s	0.32 s		Settling time	>0.50 s	0.25 s	0.30 s
Active power	Overshoot	0.00 pu	1.20 pu	1.10 pu	Active power	Overshoot	1.20 pu	1.40 pu	1.20 pu
	Dip	0.00 pu	0.00 pu	−0.20 pu		Dip	0.40 pu	0.80 pu	0.80 pu
	Settling time	0.40 s	0.25 s	0.22 s		Settling time	>0.60 s	0.450 s	0.45 s
Reactive power	Overshoot	0.20 pu	0.10 pu	0.25 pu	Reactive power	Overshoot	0.75 pu	0.75 pu	0.30 pu
	Dip	0.00 pu	0.10 pu	0.00 pu		Dip	−0.75 pu	−0.75 pu	−0.75 pu
	Settling time	0.25 s	0.25 s	0.25 s		Settling time	>1.00 s	<0.90 s	<0.90 s
Terminal voltage	Overshoot	0.02 pu	0.00 pu	0.00 pu	Terminal voltage	Overshoot	1.10 pu	1.01 pu	0.00 pu
	Dip	0.10 pu	0.10 pu	0.10 pu		Dip	0.15 pu	0.30 pu	0.20 pu
	Settling time	0.30 s	0.20 s	0.20 s		Settling time	>0.40 s	0.40 s	0.40 s

TABLE 11.4
Numerical Index Performance of the Wind Turbines Considering CBFCL

	DFIG wind turbine				PMSG wind turbine		
Variable	Metric	No control	CBFCL	Variable	Metric	No control	CBFCL
DC-link voltage	Overshoot	1.10 pu	1.10 pu	DC-link voltage	Overshoot	1.30 pu	1.10 pu
	Dip	0.01 pu	0.30 pu		Dip	0.40 pu	0.80 pu
	Settling time	0.40 s	0.22 s		Settling time	>0.50 s	0.35 s
Reactive power	Overshoot	0.20 pu	0.22 pu	Active power	Overshoot	1.20 pu	1.20 pu
	Dip	0.00 pu	0.11 pu		Dip	0.40 pu	0.70 pu
	Settling time	0.25 s	0.25 s		Settling time	>0.60 s	0.45 s
Terminal voltage	Overshoot	0.02 pu	0.00 pu	Terminal voltage	Overshoot	1.20 pu	1.00 pu
	Dip	0.10 pu	0.10 pu		Dip	0.40 pu	0.20 pu
	Settling time	0.30 s	0.60 s		Settling time	>0.60 s	0.40 s

REFERENCES

[1] Wind-Fuels and Technologies, International Energy Agency, 2021.

[2] A. Nduwamungu, E. Ntagwirumugara, F. Mulolani, and W. Bashir, "Fault ride through capability analysis (FRT) in wind power plants with doubly fed induction generators for smart grid technologies," *Energies*, vol. 13, no. 4260, pp. 1–26, 2020, doi: 10.3390/en13164260

[3] R. Sitharthan, M. Karthikeyan, D. S. Sundar, and S. Rajasekaran, Adaptive hybrid intelligent MPPT controller to approximate effectual wind speed and optimal rotor speed of variable speed wind turbine, *ISA Transactions*, vol. 96, pp. 479–489, (2020).

[4] K.E. Okedu, S. M. Muyeen, R. Takahashi, and J. Tamura, "Protection schemes for DFIG considering rotor current and DC-link voltage," *24th IEEE-ICEMS (International Conference on Electrical Machines and System)*, Beijing, China, August 2011, pp. 1–6.

[5] J. Ouyang, T. Tang, J. Yao, and M. Li, "Active voltage control for DFIG-based wind farm integrated power system by coordinating active and reactive powers under wind speed variations," *IEEE Transactions on Energy Conversion*, vol. 34, pp. 1504–1511, 2019.

[6] K. E. Okedu, *Introductory Chapter of the book Power System Stability*, United Kingdom: INTECH, February, 2019, pp. 1–10.

[7] H. Shao, X. Cai, Z. Li, D. Zhou, S. Sun, L. Guo, and F. Rao, "Stability enhancement and direct speed control of DFIG inertia emulation control strategy," *IEEE Access*, vol. 7, pp. 120089–120105, 2019.

[8] X. He, X. Fang, and J. Yu, "Distributed energy management strategy for reaching cost-driven optimal operation integrated with wind forecasting in multimicrogrids system," *IEEE Transactions on Systems, Man, Cybernetics: Systems*, vol. 49, pp. 1643–1651, 2019.

[9] H. W. Qazi, P. Wall, M. Val Escudero, C. Carville, N. Cunniffe, and J. O'Sullivan, "Impacts of fault ride through behavior of wind farms on a low inertia system", *IEEE Transactions on Power Systems (early access)*, June 2020, pp. 1–1, doi: 10.1109/TPWRS.2020.3003470.

[10] L. Chunli, T. Zefu, Q. Huang, W. Nie, and J. Yao, "Lifetime evaluation of IGBT module in DFIG considering wind turbulence and nonlinear damage accumulation effect," *2019 IEEE 2nd International Conference on Automation, Electronics and Electrical Engineering (AUTEEE)*, 22–24 November 2019, Shenyang, China.

[11] K.E. Okedu, S. M. Muyeen, R. Takahashi, and J. Tamura, "Use of supplementary rotor current control in DFIG to augment fault ride through of wind farm as per grid requirement," *37th Annual Conference of IEEE Industrial Electronics Society (IECON 2011)*, Melbourne, Australia, 7–10 November 2011.

[12] K.E. Okedu, S. M. Muyeen, R. Takahashi and J. Tamura, "Improvement of fault ride through capability of wind farm using DFIG considering SDBR," *14th European Conference of Power Electronics EPE*, Birmingham, United Kingdom, August 2011, pp. 1–10.

[13] K. E. Okedu, "Enhancing DFIG wind turbine during three-phase fault using parallel interleaved converters and dynamic resistor," *IET Renewable Power Generation*, vol. 10, no. 6, pp. 1211–1219, 2016.

[14] K.E. Okedu, S.M. Muyeen, R. Takahashi, and J. Tamura, "Participation of FACTS in stabilizing DFIG with Crowbar during grid fault based on grid codes," *6th IEEE-GCC Conference and Exhibition*, Dubai, UAE, February, 2011, pp. 365–368.

[15] B. Boujoudi, E. Kheddioui, N. Machkour, A. Achalhi, and M. Bezza, "Comparative study between different types of control of the wind turbine in case of voltage dips," *Proceedings of the 2018 Renewable Energies, Power Systems & Green Inclusive Economy (REPS-GIE)*, Casablanca, Morocco, vol. 23–24, April 2018, pp. 1–5.

[16] K.E. Okedu, S.M. Muyeen, R. Takahashi, and J. Tamura, "Comparative study between two protection schemes for DFIG-based wind generator," *23rd IEEE-ICEMS (International Conference on Electrical Machines and Systems)*, Seoul, South Korea, October 2010, pp. 62–67.

[17] K. E. Okedu, S. M. Muyeen, R. Takahashi, and J. Tamura, "Stabilization of wind farms by DFIG-based variable speed wind generators," *International Conference of Electrical Machines and Systems (ICEMS)*, Seoul, South Korea, 10–13 October 2010, pp. 464–469.

[18] Y. Bekakra, and D. B. Attous, "Sliding mode controls of active and reactive power of a DFIG with MPPT for variable speed wind energy conversion," *Australian Journal of Basic and Applied Sciences*, vol. 5, no. 12, pp. 2274–2286, 2011.

[19] D. M. Ali, K. Jemli, M. Jemli, and M. Gossa, "Doubly fed induction generator, with crowbar system under micro-interruptions fault", *International Journal on Electrical Engineering and Informatics*, vol. 2, no. 3, pp. 216–231, 2010.

[20] D. B. Suthar, "Wind energy integration for DFIG based wind turbine fault ride through", *Indian Journal of Applied Research*, vol. 4, no. 5, pp. 216–220, 2014.

[21] M.T. Lamchich, and N. Lachguer Matlab, "Simulink as simulation tool for wind generation systems based on doubly fed induction machines," *MATLAB- A Fundamental Tool for Scientific Computing and Engineering Applications*, vol. 2, Chapter 7, INTECH Publishing, 2012, pp. 139–160.

[22] A. Noubrik, L. Chrifi-Alaoui, P. Bussy, and A. Benchaib, "Analysis and simulation of a 1.5MVA doubly fed wind power in Matlab Sim PowerSystems using Crowbar during power systems disturbances," *IEEE-2011 International Conference on Communications, Computing and Control Applications (CCCA)*, Hammamet, Tunisia, 2011.

[23] M. Nasiri, and R. Mohammadi, "Peak current limitation for grid-side inverter by limited active power in PMSG-based wind turbines during different grid faults," *IEEE Transactions on Sustainable Energy*, vol. 8, pp. 3–12, 2017.

[24] A. Gencer, "Analysis and control of fault ride through capability improvement PMSG based on WECS using active crowbar system during different fault conditions," *Elektronika ir Elektrotechnika*, vol. 24, pp. 64–69, 2018.

[25] D. M. Yehia, D. A. Mansour, and W. Yuan, "Fault ride-through enhancement of PMSG wind turbines with DC microgrids using resistive-type SFCL," *IEEE Transactions on Applied Superconductivity*, vol. 28, pp. 1–5, 2018.

[26] M. Islam, M. Huda, J. Hasan, M. A. H. Sadi, A. AbuHussein, T. K. Roy, M. Mahmud, "Fault ride through capability improvement of DFIG based wind farm using nonlinear controller based bridge-type flux coupling non-superconducting fault current limiter," *Energies*, vol. 13, no. 7, pp. 1696, 2020.

[27] M. Firouzi, "Low-voltage ride-through (LVRT) capability enhancement of DFIG-based wind farm by using bridge-type superconducting fault current limiter (BTSFCL)," *Journal of Power Technologies*, vol. 99, no. 4, pp. 245–253, 2020.

[28] M. R. Islam, J. Hasan, M. M. Hasan, M. N. Huda, M. A. H. Sadi, and A. AbuHussein, "Performance improvement of DFIG-based wind farms using narma-12 controlled bridge type flux coupling non-superconducting fault current limiter," *IET Generation, Transmission & Distribution*, vol. 14, no. 26, pp. 6580–6593, 2021.

[29] J. Hasan, M. R. Islam, M. R. Islam, A. Z. Kouzani, and M. P. Mahmud, "A capacitive bridge-type superconducting fault current limiter to improve the transient performance of DFIG/PV/SG-based hybrid power system," *IEEE Transactions on Applied Superconductivity*, 2021, doi: 10.1109/TASC.2021.3094422.

[30] G. Rashid, and M. H. Ali, "A modified bridge-type fault current limiter for fault ride-through capacity enhancement of fixed speed wind generator," *IEEE Transactions on Energy Conversion*, vol. 29, no. 2, pp. 527–534, 2014.

[31] M. M. Moghimian, M. Radmehr, and M. Firouzi, "Series resonance fault current limiter (SRFCL) with MOV for LVRT enhancement in DFIG-based wind farms," *Electric Power Components and Systems*, vol. 47, no. 19–20, pp. 1814–1825, 2019.

[32] J. Yang, J. E. Fletcher, and J. O'Reilly, "A series-dynamic-resistor-based converter protection scheme for doubly-fed induction generator during various fault conditions," *IEEE Transactions on Energy Conversion*, vol. 25, no. 2, pp. 422–432, 2010.

[33] Z. Din, J. Zhang, Z. Xu, Y. Zhang, and J. Zhao, "Low voltage and high voltage ridethrough technologies for doubly fed induction generator system: Comprehensive review and future trends," *IET Renewable Power Generation*, vol. 15, no. 3, 614–630, 2021.

[34] M. Firouzi, and G. Gharehpetian, "Improving fault ride-through capability of fixedspeed wind turbine by using bridge-type fault current limiter," *IEEE Transactions on Energy Conversion*, vol. 28, no. 2, pp. 361–369, 2013.

[35] S. B. Naderi, M. Jafari, and M. T. Hagh, "Parallel-resonance-type fault current limiter," *IEEE Transactions on Industrial Electronics*, vol. 60, no. 7, 2538–2546, 2012.

[36] PSCAD/EMTDC Manual, Manitoba HVDC lab, 2016.

[37] D. Yang, X. Ruan, and H. Wu, "Impedance shaping of the grid-connected inverter with LCL filter to improve its adaptability to the weak grid condition," *IEEE Transactions on Power Electronics*, vol. 29, no. 11, pp. 5795–5805, 2014.

[38] C. V. Thierry, *Voltage Stability of Electric Power Systems*. Springer, 1998.

[39] S. Grunau, and W. F. Fuchs. "Effect of wind-energy power injection into weak grids," *Institute for Power Electronics and Electrical Drives, Christian-Albrechts-University of Kiel D-24143*, Kiel, Germany, pp. 1–7, 2012.

[40] M. R. Islam, M. N. Huda, J. Hasan, M. A. H. Sadi, A. Abuhussein, T. K. Roy, and M. A. Mahmud, "Fault ride through capability improvement of DFIG based wind farm using nonlinear controller based bridge-type flux coupling non-superconducting fault current limiter", *Energies*, vol.13, p. 1696, 2020, doi: 10.3390/en13071696.

[41] G. Rashid, and M. H. Ali, "Fault ride through capability improvement of DFIG based wind farm by fuzzy logic controlled parallel resonance fault current limiter," *Electric Power Systems Research*, vol. 146, pp. 1–8, 2017.

[42] G. Rashid, and M. H. Ali, "Nonlinear control-based modified BFCL for LVRT capacity enhancement of DFIG-based wind farm," *IEEE Transactions on Energy Conversion*, vol. 32, pp. 284–295, 2016.

[43] K. E. Okedu, "Augmentation of DFIG and PMSG Wind Turbines Transient Performance Using Different Fault Current Limiters," *Energies* 2022, 15, 04817, pp. 1–25, doi. org/10.3390/en15134817.

12 DFIG and PMSG Wind Turbines Life Cycle Cost Analysis

12.1 CHAPTER INTRODUCTION

Despite the proliferation of wind farms and the high penetration of wind turbines in existing power grids, the maintenance costs for wind turbines are usually high, and this would hinder the wide utilization of wind power to help achieve the target of wind turbine installations [1]. The frequent occurrence of faults in the DFIG and PMSG wind turbines would cause unexpected downtimes leading to high maintenance cost and production losses [2]. Consequently, there is need for optimal maintenance topologies and decisions in order to cut down costs [3]. The employment of Life Cycle Cost (LCC) analysis could help in this regard as a fundamental tool to achieve a more cost-effective maintenance strategy for the DFIG and PMSG wind turbines. This would definitely boost wind power and create a competitive electricity price. In order to mitigate maintenance costs in DFIG and PMSG wind turbines, Condition Monitoring Systems (CMS) may be a reliable tool. CMS is commonly employed in industries to constantly monitor the performance of the generator, gearbox and transformer of wind turbines and help predict best time for maintenance. Most wind turbine manufacturing companies are developing and testing CMS [4], by carrying out vibrations analysis, especially for DFIG and PMSG wind turbine wheels, bearings in the gearbox and generator, along with the use of microprocessor-based systems to automate lubricating oil analysis in the gearbox and drive train system of the wind turbines. Based on this, it is easier to know the best and most cost-effective period for oil change [5]. Reliability-centered Maintenance (RCM) is a further step of CMS, with focus on the reliability aspects during maintenance. This is done by giving priority to critical components in the wind turbines with regard to maintenance tasks [6].

12.2 FUNDAMENTAL AND THEORETICAL BACKGROUND

In this section, some fundamental and theoretical terminologies would be discussed with regard to the concept of LCC analysis.

12.2.1 RELIABILITY, AVAILABILITY AND DOWNTIME

Reliability could be defined as "the ability of an item to perform a required function under given conditions for a given time interval" [7], without fail, while availability is the "ability to be in a state to perform as and when required, under given conditions, assuming that the necessary external resources are provided" [7].

DOI: 10.1201/9781003350910-12

Mathematically, the average availability of any device or equipment could be expressed as [8]:

$$A_{av} = \frac{MTTF}{MTTF + MTTR} \qquad (12.1)$$

where MTTF is the Mean Time to Failure, measured in hours [h], and this defines the average operating time of the item, while MTTR is the Mean Time to Repair, measured in hours [h], and this expresses the average of the times required to restoration after failures of the item.

Downtime is the "time interval throughout which an item is in a down state" [7]. Figure 12.1(a) and (b) shows downtimes due to replacements. Figure 12.1(a) shows the MTTR of Corrective Maintenance (CM), during actual failure, while Figure 12.1(b)

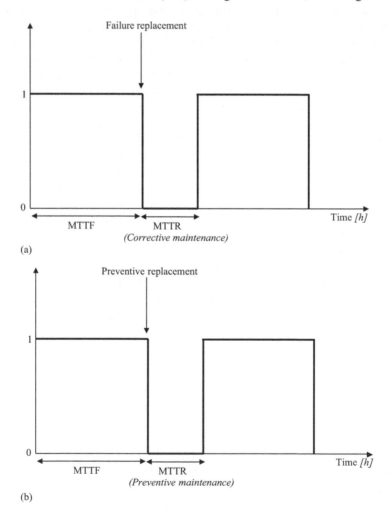

FIGURE 12.1 MTTF and downtime for different types of maintenance [9]. (a) MTTR of Corrective Maintenance (b) MTTR in case of Preventive Maintenance.

illustrates the MTTR in case of Preventive Maintenance (PM), where no actual failure occurred, rather only a preventive replacement.

12.2.2 OPERATION AND MAINTENANCE COST, FAILURE MODE AND EFFECTS ANALYSIS

The DFIG and PMSG wind turbines need regular checks during the course of their lifetime, and these associated costs are the operations and maintenance costs. Failure Mode and Effects Analysis (FMEA) is used in operations management for the analysis of failure modes in inductive failure analysis within a system, based on the severity and likelihood of the failures in a classified manner. One of the key benefits of a good FMEA is that it helps to know the potential failure modes based on previous data with the same sample space in other to reduce costs of development. Basically, the total cost of DFIG and PMSG wind turbines during their lifetime could be computed by the sums of all expenses during their lifetime. It should be noted that this total cost is divided into three categories: installation costs of the DFIG and PMSG wind turbines, capital costs of the DFIG and PMSG wind turbines and Operation and Maintenance (O&M) costs. The ratio of the total money invested in a wind farm project to the actual number of DFIG and PMSG wind turbines gives the total installation costs, while the interest on the money invested in building the DFIG and PMSG wind farms is known as the capital costs.

12.3 LIFE CYCLE COST ANALYSIS OF THE DFIG AND PMSG WIND TURBINES

The definition of the LCC concept could be expressed as [7]: "all the costs generated during the life cycle of an item." This terminology is usually used for cost saving in investments and it relates to the computational strategies of total lifetime costs of the item. Furthermore, this concept promotes comparative cost assessment over a given period of time, considering the relevant economic factors based on initial capital and future operational costs [10]. The comparison of different investment options is also possible based on the LCC concept, by estimating the actual investment costs, considering the initial investment cost and other related costs to the item's whole lifetime. Mathematically, the LCC could be expressed as [11]:

$$LCC = C_{INV} + C_{CM} + C_{PM} + C_{PL} + C_{REM} \qquad (12.2)$$

where C_{INV} is the input cost of investment, C_{CM} is the cost for CM, C_{PM} is the cost PM, C_{PL} is the cost for production loss and C_{REM} is the remainder value.

12.3.1 PRODUCTION LOSSES OF THE DFIG AND PMSG WIND TURBINES

In the course of the lifetime of the DFIG and PMSG wind turbines, failures may occur and this would lead to unexpected downtimes. In light of this, it is possible to determine the production losses cost by using the following expression [12]:

$$C_{PL} = N.P.C_f.C_{el}.D \qquad (12.3)$$

where N is the number of DFIG or PMSG wind turbines, P is the generated electric power, C_f is the capacity factor, which is defined as the ratio of the actual power generated to the maximum power that could be generated, expressed as:

$$C_f = \frac{P_{out}}{P_{max}} \qquad (12.4)$$

C_{el} is the cost of electricity and D is the downtime expressed in hours.

12.3.2 Preventive Maintenance of the DFIG and PMSG Wind Turbines

Any maintenance that is undertaken before failures happen is known as PM. It is classified into two: preventive scheduled maintenance, which is the PM done in accordance with an established time schedule [13] and preventive condition-based maintenance, which is the PM done based on performance and parameter monitoring and subsequent actions. For accurate prediction of when maintenance is required, the history of when, how and why the component has failed is important [11]. The good motivation for any DFIG and PMSG wind turbines PM topology is that of reduced cost of PM measures lower than no action [14].

12.3.3 Corrective Maintenance of DFIG and PMSG Wind Turbines

The maintenance done after fault recognition with the intention to put an item into a state in which it can perform a required function is known as CM [13]. This means that no maintenance is required as long as the item is working, however, the item is fixed or repaired or removed when is not working anymore. The scenario where little or no PM is carried out would result in more system failures requiring more CM. The PV method can be used to compare future payments over a certain time at one point in the present time. The Present Value (PV) of the item is the current amount of the item at a certain rate (d) to pay for an outlay (C) after n years. This implies that all future payments are recomputed to the equivalent value for the present time. The PV of one outlay (C) to be paid after n number of years is gained by the product of this value and the PV factor $PV_f(n,d)$ expressed as [15]:

$$PV = C.PV_f(n,d) = C.(1+d)^{-n} \qquad (12.5)$$

Figure 12.2 shows the different maintenance strategies that could be applied to the DFIG and PMSG wind turbines, while Figure 12.3 displays the component conditions based on different strategies.

12.3.4 Reliability-Centered Maintenance

A more structured approach that concerns reliability aspects during maintenance plans of an item is known as RCM [16]. This concept would help in balancing PM and CM and would enhance the best PM activities to be carried out for the right components of the wind turbines at the right time to achieve the most efficient solutions

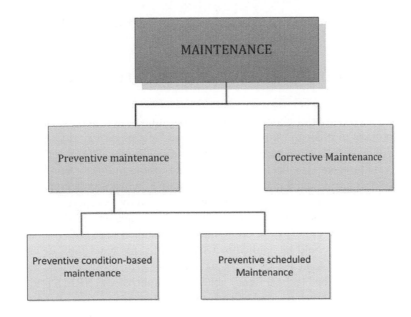

FIGURE 12.2 Different maintenance strategies [13].

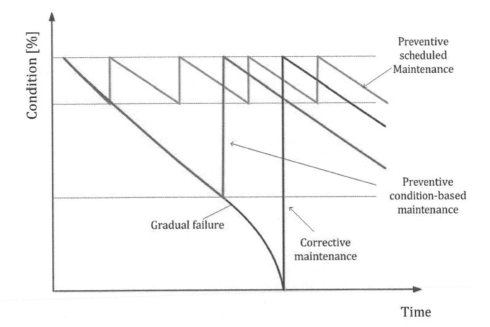

FIGURE 12.3 The conditions of the components based on the different maintenance strategies [9].

[15]. Thus, RCM could be termed as one strategy to improve the reliability of physical assets.

12.4 TYPICAL LEVELIZED COST OF ENERGY FOR WIND FARM PROJECTS

This section gives a brief about representative of utility-scale and distributed wind energy projects in order to estimate the Levelized Cost of Energy (LCOE) for both land-based and offshore wind power plants that could be made up of DFIG and PMSG wind turbines. The data and the presented results in this section were based on the 2019 commissioned plants and representative industry data considering the state-of-the-art modeling capabilities in the US case study as reported in [17]. The modeling carried out is geared toward more granular details on specific cost categories. The case study in [17] is based on the 2019 installed land-based wind project data and costs that could be applied to wind turbines for wind farm development and are primarily obtained from Wiser and Bolinger [18]. However, the data were supplemented with the National Renewable Energy Laboratory (NREL) [19] outputs cost models for wind turbines and balance-of-system details.

This section estimates the LCOE for land-based and offshore wind projects. LCOE is a metric employed to assess the cost of generating electricity, considering the total level impact of the plants based on technological changes in design compared to the costs of all types of generations possible.

For wind farm projects, the mathematical expression for LCOE is:

$$LCOE = \frac{(CapEx \times FCR) + OpEx}{(AEPnet / 1,000)} \qquad (12.6)$$

where
   LCOE is the Levelized Cost of Energy ($/megawatt-hour [MWh]),
   FCR is the fixed charge rate (%),
   CapEx is the capital expenditures ($/kilowatt [kW]),
   AEPnet is the net average annual energy production (MWh/megawatt [MW]/
      year [yr]) and
   OpEx is the operational expenditures ($/kW/yr).

In Equation (12.6), CapEx, OpEx and AEPnet are the first three basic inputs that enable the equation to capture system-level impacts as a result of design changes due to larger rotors or taller wind turbine towers of the DFIG and PMSG wind turbines. FCR is the fourth-basic input that gives the amount of revenue required to pay the annual carrying charges as applied to the CapEx on that investment during the expected project economic life. Figures 12.4–12.7 show the capital expenditures and LCOE expenditures for land and offshore wind reference projects considering the case of the United States [17].

Figure 12.4 shows that the Nacelle, the rotor and the tower of the DFIG or PMSG wind turbines, hold a large share of the capital expenditures for land-based wind farm

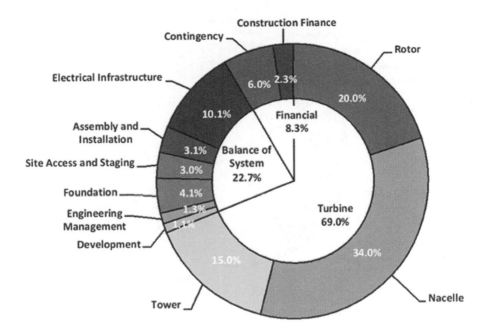

FIGURE 12.4 Capital expenditure for the land-based reference wind farm power project [17].

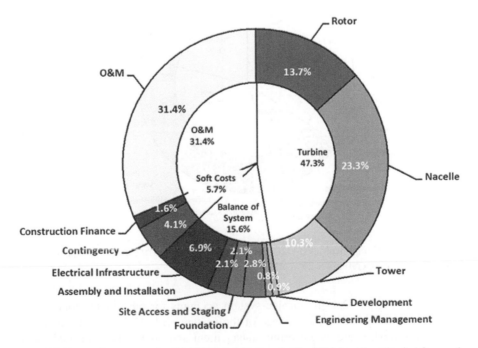

FIGURE 12.5 Component-level LCOE contribution for the 2019 land-based-wind farm reference project operating for 25 years [17].

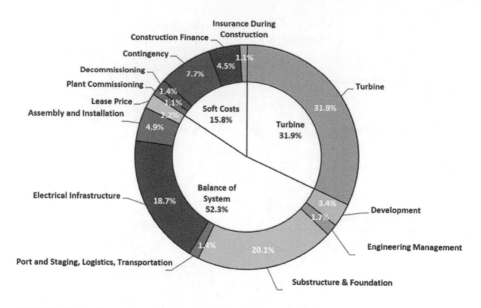

FIGURE 12.6   Capital expenditure for offshore-based reference wind farm power project [17].

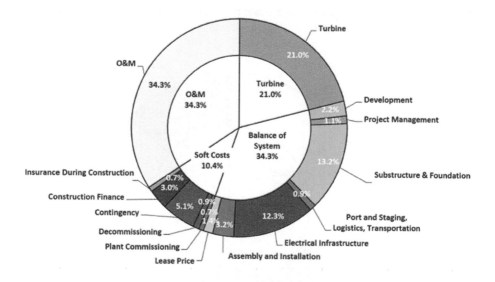

FIGURE 12.7   Component-level LCOE contribution for the 2019 offshore-based-wind farm reference project operating for 25 years [17].

projects. In Figure 12.4, the turbine system accounts for 69% of the total capital expenditure compared to the balance of the system with 22.7% which comprised electrical infrastructure, engineering management and others. The financial aspect takes 8.3% of the capital expenditure, which is basically on contingencies and financial construction. The component level for the LCOE for the land-based wind farm

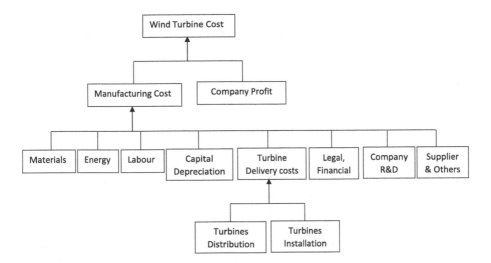

FIGURE 12.8 Wind turbine cost decomposition in its sub-categories of cost [20].

project in Figure 12.5 reflects that O&M has a larger share with 31.4%, and 23.3% for the nacelle. The turbine system is having the highest LCOE with 47.3%, compared to the balance of system, O&M and soft cost.

On the other hand, the capital expenditure for the offshore-based wind farm project in Figure 12.6 reflects that the balance of the system is holding the largest share of 52.3%, compared to the turbine system share of 31.9% and soft cost of 15.8%. However, the balance of the system and the O&M cost hold the same share of 34.3%, which is the largest share for the component level LCOE for the offshore-based wind farm projects in Figure 12.7.

Figure 12.8 shows the wind turbine cost decomposition of various sub-classes of cost expenditures. Figure 12.8 includes the entire cost of manufacturing, to build and deliver a typical wind turbine to a wind farm [20], as can be likened to DFIG and PMSG wind turbines. The manufacturing cost seems to be high since it is comprised of the materials to be used for the wind turbines, energy, labor, capital depreciation, delivery cost, legal financial aspects of the wind turbines, Research and Development (R&D) phase, supplier and others. Furthermore, the turbine delivery cost consists of two aspects: turbine distribution and turbine installation. All wind farm developers and policy makers follow the outlines of Figure 12.8.

12.5 CHAPTER CONCLUSION

In this chapter, some of the LCC strategies that could be applied to DFIG and PMSG wind turbines, for effective-cost reduction, were presented. The fundamental and theoretical background of LCC were discussed with regard to reliability, availability and downtime. The chapter also tackled O&M cost that may arise in the course of operation of these wind turbines considering to avoid FMEA. A typical LCC analysis that could be linked to DFIG and PMSG wind turbines in order to reduce production losses and enhance PM and CM was also reported. Furthermore, an LCOE for both land and offshore wind farm projects using a case study was also reported in this chapter. Finally, a typical wind turbine cost decomposition from manufacturing to delivery was also presented.

PERSPECTIVE

The technology of variable speed wind turbines is very promising in renewable power generation. It is imperative for wind turbines to gain control after grid disturbances and contribute to the stability of power grids, as part of the requirements of grid codes set by grid operators in operating wind farms. The Doubly Fed Induction Generator (DFIG) and the Permanent Magnet Synchronous Generator (PMSG) are the two most commonly used variable speed wind turbines in modern wind farms. However, both wind turbines are affected during transient state, when there is disturbance in the grid. A better terminology for this grid disturbance is known as the Low Voltage Ride Through (LVRT) or Fault Ride Through (FRT) capabilities of the wind turbines. Many suggestions were given in the literature to solve the LVRT or FRT issues for both wind turbines.

Fault Current Limiters (FCLs) are capable of augmenting the performance of wind turbines during grid disturbances. In this book, the augmentation of the DFIG and the PMSG wind turbines was robustly studied, considering various levels and topologies of operation, using FCLs. The considered FCLs were the Series Dynamic Braking Resistor (SDBR), Bridge Fault Current Limiter (BFCL) and the Capacitive Bridge Fault Current Limiter (CBFCL). The wind turbines were operated under the same conditions. Both wind turbines were affected by the connection of the FCLs in their stators during severe grid disturbance. However, the influence of the FCLs was more observed in the PMSG wind turbine compared to the DFIG wind turbine. This is due to the fact that the PMSG wind turbine technology enables the decoupling of its full back-to-back power converters at the power grid.

Some other FRT strategies like super capacitor system for the DFIG and the effects of the machine parameters for both DFIG and PMSG were also studied in this book. Efforts were made to use only the effective values of the machine parameters and FCLs in the evaluation of the system. In addition, the cost analysis of both wind turbines was reported, considering a case study for practical realization.

It could be concluded that, while the DFIG wind turbine is cheaper and widely used in wind farms development and wind power applications, the PMSG performs better during transient state due to the technology of fully decoupling its active and reactive power in the power network, since it is fully rated. Therefore, it is recommended to use FCLs with DFIG- or PMSG-based variable speed wind turbine to get superior LVRT or FRT performance, based on the intended purpose of the power grid operator or energy policy makers.

REFERENCES

[1] G. Puglia, "Life cycle cost analysis on wind turbines," Master of Science Thesis in Energetic Engineering, Department of Energy and Environment, Chalmers University of Technology, Gothenburg, Sweden, pp. 1–73, 2013.

[2] K. Fischer, F. Besnard, and L. Bertling, *A Limited-Scope Reliability-Centered Maintenance Analysis of Wind Turbine*," Brussels: EWEA, 2011.

[3] F. Besnard, and L. Bertling, "An approach for condition-based maintenance optimization applied to wind turbine blades," *IEEE Transactions on Sustainable Energy*, vol. 1, no. 2, pp. 77–83, 2010.

[4] J. Nilsson, "Maintenance management of wind power systems-Cost effect analysis of conditioning monitoring system," Master's thesis, KTH, Stockholm, 2006.

[5] A. Davies, *Handbook of condition monitoring: Techniques and Methodology*, London: Chapman & Hall, 1998.

[6] O. Wilhelmsson, "Evaluation of the introduction of RCM for hydro power generators at Vattenfall Vattenkraft," Stockholm, Master thesis KTH, XR-EE-EEK 2006:009, 2005/2006.

[7] Swedish Standard Institute, "EN13306," 2010.

[8] M. Rausand, and A. Høyland, *System Reliability Theory*, Hoboken: John Wiley & Sons, 2004.

[9] A. K. S. Jardine, and A. H. C. Tsang, *Maintenance, Replacement, and Reliability*, USA: Taylor & Francis, 2006.

[10] P. Gluch, and H. Baumann, "The life cycle costing (LCC) approach: a conceptual discussion of its usefulness for environmental decision-making," *Building and Environment*, vol. 39, pp. 571–580, 2004, doi:10.1016/j.buildenv.2003.10.008.

[11] J. Nilsson, and L. Bertling, "Maintenance management of wind power systems using condition monitoring systems-life cycle cost analysis for two case studies," *IEEE Transactions on Energy Conversion*, vol. 22, no. 1, pp. 223–229, 2007.

[12] Interview with F. Besnard, PhD Student, Chalmers University, Göteborg, Sweden.

[13] IEC 60050-191, International Vocabulary (IEV) Chapter 191: Dependability and quality of Service, International Electrotechnical Commission (IEC), 2013.

[14] L. Bertling, R. Allan, and R. Eriksson, "A reliability-centered asset maintenance method for assessing the impact of maintenance in power distribution system," *IEEE Transactions of Power System*, vol. 20, no. 1, pp. 75–82, 2005.

[15] L. Bertling, F. Besnard, and J. Nilsson, *On the Economic Benefits of Using Condition Monitoring Systems for Maintenance Management of Wind Power Systems*, Stockholm: IEEE, 2010.

[16] D. McMillan, and G. Ault, *Towards Quantification of Condition Monitoring Benefit for Wind Turbine Generators*, UK: Institute for Energy & Enviroment, University of Strathclyde, 2007.

[17] T. Stehly, P. Beiter, and P. Duffy. *Technical Report on 2019 Cost of Wind Energy Review*, National Renewable Energy Laboratory, pp. 1–86, 2020.

[18] R. Wiser, and M. Bolinger, 2016. *2015 Wind Technologies Market Report*. Washington, DC: U.S. Department of Energy, Office of Energy Efficiency and Renewable Energy. http://energy.gov/sites/prod/files/2016/08/f33/2015-Wind-Technologies-Market-Report08162016.pdf

[19] National Renewable Energy Laboratory (NREL), "Wind integration national dataset toolkit." https://www.nrel.gov/grid/wind-toolkit.html (accessed on December 2020).

[20] A. Elia, M. Taylor, B. O. Gallachoir, and F. Rogan, "Wind turbine cost reduction: A detailed bottom-up analysis of innovation drivers," *Energy Policy*, vol. 147, no. 111912, pp. 1–30, 2020.

Index